According to the Map

According to the Map

deciphering an extraterrestrial message

Sterling Kodiak

Copyright @2018 by Sterling Kodiak

All rights reserved. No part of this book may be used or reproduced, stored in a retrieval system, or transmitted in any form or by any means, electronic, mechanical, photocopying, recording, or otherwise without prior written permission of the copyright owner.

ISBN 978-0-9998122-0-4
Library of Congress Control Number 2018901507

Illustration from "The Case For A Plan Centred On Khafre" reprinted by permission.

Illustrations of the Giza Vanishing Point by John Legon reprinted by permission.

Cover Photography by Steven Allan, IStockPhoto

Research and methods presented in this book reflect the author's personal experiences and thoughts. Although the author and publisher have made every effort to ensure that information in this book is correct at press time, the author and publisher do not assume and hereby disclaim any liability to any party for any loss, damage, or disruption caused by errors or omissions, whether such errors or omissions result from negligence, accident, or any other cause.

Printing/manufacturing information for this book may be found on the last page.

Published by Webscrolls Publishing
www.webscrolls.com

Email: info@quantumsubjectivity.com

We have nothing to fear but ourselves

Contents

	Preface	ix
	Introduction	xiii
1	**The Search for Meaning**	1
	Introduction	
	From the Obvious to the Not So Obvious	
	The Map	
	An Interpretation	
	Dimensions of the Situation	
	Where We Are and Where We're Headed	
2	**Perception of (an Object)**	17
	Introduction	
	Understanding the Situation	
	Levels of Abstraction	
	Where the Pyramids Fit In	
	The Progression of States	
	Frames of Reference	
	The Uncertainty of Observation	
	Back to the Pyramids	
	What's Missing	
	The Inner and Outer Horizons	
	The Certainty Frame of Reference of Perception	
	Conceptualizing Consciousness	
	Expectation and Memory	
	Perception as a Process	

3 Consciousness of (Perception of (an Object)) 47
Introduction
The Vanishing Point
Subjective Perception of (an Object)
Consciousness of (Perception of (an Object))
The Collapse to Certainty
Conceptualizing Perception of (an Object)
A Moment of Reflection about a Moment
Tell Tale Patterns
Let's Take Our Bearings
Sidelights
The Great Pyramid and its Chambers
What's Cosmic About It
Summary

4 Know Thyself 97
Take a Ride with Me

References   107
Appendix 1   111
Appendix 2   112
Appendix 3   115
Appendix 4   121
Appendix 5   127
Notes   131
Epilogue   175

Preface

Are the pyramids of Giza an encoded message? For an encoded message to be of value, it needs to be decoded. It also helps to know who authored the message. Perhaps some clarity can be found with the aid of a map.

The design of the pyramids encodes a map presenting a generic situation that can be interpreted as specific situations capable of providing insight into what makes us the beings we are. Following the map requires interpreting the generic situation as specific situations in phenomenology, psychology, and ontology. In this presentation we'll follow the map via a high level interpretation of the phenomenological situation.

The pyramids as map proposal is less challenging to fathom than a complementary proposal: the map is built on three progressive levels of abstraction forming a conceptual geometry consistent with a multidimensional cosmos and its temporal dynamics–physics that eludes us to this day. Assuming we can establish the validity of these proposals, what are they telling us?

If the design of an ancient monument can be interpreted as the representation of specific situations built on a conceptual geometry consistent with fundamental physics we don't fully understand, then the monument is arguably of extraterrestrial design. According to the map, we're not alone in the universe. (Don't confuse three levels of abstraction with the three specific situations mentioned above. The highest level of abstraction will be referred to as a conceptual framework. A conceptual framework can be developed for each specific situation of interest.[1])

We're not suggesting someone other than ancient Egyptians built the pyramids, nor are we questioning the ability of an ancient civilization to accumulate sufficient knowledge and experience required to engineer a structure as spectacular as the Great Pyramid. What's suspect is the assumption that ancient Egyptians accumulated enough knowledge about cosmic multidimensionality, temporal dynamics, and organismic development to enable them to encode a map—accomplished by combining all three bodies of knowledge—for the purpose of inviting interpretations of the map as insight into human nature, which means the designer of the map must have known about interpretations of specific situations in phenomenology, psychology, and ontology representing interrelated articulations of a generic situation. Needless to say, the knowledge requirements would have been a tall order for an ancient civilization.

Our goals are to turn the hypothesis of an extraterrestrial message into a conclusion and highlight research that may be the key to completely understanding what Earth's visitors were trying to tell us. To achieve these goals the argument for extraterrestrial visitation is presented at a level called popular science. This level of presentation requires abstractions that may take some getting used to, but anyone mildly familiar with object-oriented thinking required in computer science, the kind of graphic analysis popular in economics, or reflections of the type common in philosophy should find the material easy to assimilate.[2]

The first three chapters of the book are dedicated to interpreting the generic situation represented by the pyramids as a phenomenological situation: Chapter one starts by outlining an approach that allows a generic situation we observe to be interpreted as a specific situation we understand. Chapter two uses the approach of chapter one and the second level of abstraction to interpret the generic situation as perception of (an object)—a good start but not a complete interpretation. Consciousness is introduced in chapter three to complete the interpretation of the generic situation as a phenomenological situation we can describe as consciousness of (perception of (an object)). In chapter three, objective consciousness is presented as the actualization of potential represented by subjective perception of (an object) contingent on the object being observed as it appears in the now. The chapter concludes with additional insight into the relationship between consciousness and perception provided by the Great Pyramid. The Great Pyramid brings the presentation full circle by revealing compelling evidence supporting the interpretation of the pyramid complex as a map. By the end of chapter three, it will be hard to deny the design of the pyramids could be extraterrestrial in origin.

Finally, the author uses chapter four to recollect how the seeds of research responsible for the sidelight this book represents were sown. It's a personal perspective.

No matter how much evidence we provide, some will have difficulty accepting the design of the pyramids as the encoding of a message of extraterrestrial origin. The reasons for this reluctance range from understandable to disappointing. Nevertheless, we have a dilemma: suggesting an ancient civilization achieved the insight required to encode state of the art knowledge in the manner we propose is as hard to believe as suggesting an extraterrestrial source of that knowledge.

The one thing that's certain is the map didn't encode itself. The knowledge encoded in the design of the pyramids didn't appear out of nowhere. Why was it so important to insure an encoding of knowledge would survive for centuries if the designer wasn't confident about the value of that knowledge? Is it a coincidence that the knowledge provided is knowledge we currently seek? Before we can concern ourselves with who designed the map, we need to concern ourselves with what it's telling us. A better understanding of what the pyramids map and why the map was created might provide a better understanding of who designed it.

(The insights presented in this book are based on discoveries supporting research in artificial consciousness already documented in significant detail: a document that will be referred to as the *primary document*.[3] In the primary document there's no mention of pyramids, maps, or even a hint of the possibility of extraterrestrial contact. There's only the usual barrage of equations, diagrams, and terminology directed at a more specialized audience.[4])

Introduction

Suppose you and I, members of a team of intergalactic explorers, discover alien humanoids on a distant planet—beings that can be described, from our technological perspective, as little more than a few steps out of their Stone Age. How much interaction, if any, is appropriate? Do we subscribe to the idea that primitive beings should be allowed to experience the ups and downs that come with developing their own civilization, or do we help them develop someone else's vision of how a civilization should evolve? What if the humanoids tend to be hostile or misinterpret our technology and start worshiping us as gods? In whose interest do we act?

All too soon the exuberance that comes from satisfying our curiosity is tempered by the motives of those who fund our efforts. Exploration to discover mineral resources or a planet to colonize is subject to different guidelines than exploration to discover new life-forms. Different guidelines for exploration suggest different priorities when it comes to the well-being of indigenous life. To appreciate a scenario where we show empathy for the humanoids, let's start by introducing a scenario where we don't.

In the minds of primitive humanoids, our arrival is akin to gods descending from the heavens. It's a fascination destined to be short lived. An Earth-like planet, rich in resources and already supporting complex life, invites a rush of earthlings claiming land and mining minerals. That rush doesn't bode well for the humanoids: we invariably treat them as an impediment or find a way to exploit them. Even the most primitive of beings soon see us for what we are and long for us to ascend back into the heavens from which we came, and if for some reason we find it in our best interest to abandon the planet, we probably leave as abruptly as we arrived. Myths and images created by conflicted humanoids become the only record of our arrival—an intrusion best forgotten.

Now assume we neither colonize the planet nor extract its resources. We care enough about the well-being and destiny of the humanoids to recognize that our continued presence is bound to become problematic. In this scenario our altruism compels us to discover, research, and move on; but before we move on, we intend to craft a message for the humanoids' descendants.

Crafting our message requires choosing a message. One possibility is a message celebrating some great mystery of science we struggled with and solved. It's an appealing idea with a couple of considerations. When passing on scientific knowledge—any knowledge—timing becomes an issue. In order for knowledge to be of value to the beings receiving it, they have to be developed enough to assimilate that knowledge and reconcile it with their world-view. Focusing on scientific knowledge also expresses an expectation regarding the primacy of scientific advancement as a measure of progress. Another possibility is a message answering the question of what's important in life. It's a harder question to answer than it appears because the answer is a matter of opinion. Finally, we come to a message about life's place in the cosmos. Offering knowledge of life's place in the cosmos assumes considerable confidence in the integrity of our enlightenment. How do we suggest to the humanoids that their nature, like human nature, is the personification of something cosmic?

Choosing the right message is easier when we have an idea of what we expect it to accomplish. At a minimum, we want the message and its presentation to support the goal of convincing future generations of humanoids they're not alone in the universe. It's a tricky goal to achieve because an obvious message is unlikely to survive the self-interest of the rise and fall of kingdoms, politics, and religions. Given what we've managed to piece together regarding the rise and fall of great civilizations due to war and environmental cycles, given the effectiveness of the establishment at suppressing any perceived threat to its authority, and given the kinds of intolerance religion is capable of breeding, the encoding of the message into its symbolic representation needs to be an enigma that can only be decoded by a civilization with a comprehensive accumulation of knowledge and no restrictions on using that knowledge to answer fundamental questions. We want the encoding to be sufficiently intriguing and the message to be sufficiently insightful to convince future generations of humanoids they're interpreting the legacy of someone other than their ancestors. If the encoding turns out to be the enigma we intend it to be, it could take the humanoids a long time to recognize the message for what it is. The upshot is we're obliged to leave a message capable of surviving centuries of humanoid development. The message needs to be an encoding of significant permanence–something hiding in plain sight for all to contemplate.

Sometime around the dawn of human civilization, Earth could have been visited by explorers from another planet. We could be those indigenous primitives described above. Speculation that Earth was visited by extraterrestrials is difficult to resolve because the evidence tends to be anecdotal and not completely convincing. What's needed is decisive evidence, a message of sophistication extraterrestrial explorers left behind to demonstrate they were here. If the message exists and can be found, it changes our entire view of life in the cosmos.

So where on planet Earth do we look for such a message? Unless the message is someplace undiscovered, we want to look at the usual candidates capable of encoding knowledge in a way currently defying interpretation. One candidate always stands out: the Giza pyramid complex, located on the Giza Plateau near Cairo, is arguably the most engaging monument ever built.

Why is curiosity about the purpose of the pyramids unbridled by arguments promoting a pharaoh's tomb? The archaeological narrative focusing on the Great Pyramid reconciles its construction with our understanding of ancient history, but it's a narrative built on a tomb interpretation that isn't a certainty. The intended occupant of the structure might have been symbolic. A pyramid constructed to be nothing more than a tomb should have walls covered with hieroglyphics extolling the greatness of the pharaoh and his accomplishments, yet what we see in the Great Pyramid is a noted absence of hieroglyphics that can only be by design. When interpreted from the intended perspective, the pyramids are presented without hieroglyphics because they're fundamental in the sense of supporting different conceptualizations. The pyramids were designed to invite us to reflect on what they represent. The effectiveness of that design is demonstrated by our continued fascination with discovering their purpose.

Could the monument have been designed to serve the dual purpose of a message and a pharaoh's tomb? Maybe. Is a tomb without a verified occupant the consequence of circumstances, or is it by design? With no verified occupant of the Great Pyramid, our attention can only be directed toward the pyramid itself. So, from the perspective of keeping us focused on what the pyramids represent, a tomb without an occupant is understandable.

Construction of the pyramids would have been a massive jobs project. If job creation was the motive, why weren't resources dedicated to some kind of infrastructure, commerce, or defense effort with a tangible return? Even if we appeal to the need for the pharaoh to project his authority, there might have been a more manageable project to satisfy that need. The investment required to build the pyramids becomes easier to understand if the pyramids were intended to be more than a tomb. Memorializing the message and its authorship was so important to someone that no expense was spared to bring the project to fruition.

No conversation about the pyramids can duplicate the astonishment you feel when you visit the site, walk up to the Great Pyramid, and gaze upward to gauge the enormity of the effort. You've probably heard something like this before, but what you don't hear about is the total bewilderment you feel after making the climb up the Grand Gallery of the Great Pyramid to find yourself standing in the King's

Chamber staring at four blank granite walls in a space that seems underwhelming by comparison to the overall structure. All of that stone, all of the man-hours of labor, all of the feats of engineering required—you just know there's more to the chamber's purpose than what you're seeing. You're thinking, "There's something else going on here."

Chapter 1

The Search for Meaning

Introduction

Knowing physical dimensions of the Great Pyramid and its chambers allows us to assume we're knowledgeable, yet speculations about the true purpose of the structure remind us of how much we don't know. The problem is we tend to forget that the Great Pyramid is the prominent structure in a pyramid complex with a gestalt we don't fully understand. Observations of how the pyramids of Giza are positioned with respect to each other and of how they align with stars in the constellation Orion can be found–they're certainly valuable contributions–but what can't be found is a convincing explanation of how all of the observations and all of the structures of the complex coalesce into an interpretation we recognize as the designer's intent.

Understanding the design of the pyramids is contingent upon interpreting what they map. The pyramids map a generic situation that needs a familiar situation to reveal their true purpose. Our goal is to interpret the pyramids from the perspective of the human situation in the context of a multidimensional environment.

From the Obvious to the Not So Obvious

Let's start by contemplating what we observe. We see three very similar looking pyramids clearly displaying a progression in size. Two ways to interpret the progression immediately come to mind: either each pyramid represents something different, suggesting the progression is meant to distinguish each represented entity from the other two, or what's communicated is the representation of one entity in three different states, suggesting each pyramid represents a different state of the same entity. Because our experience of the world is typically described using a subject-object dichotomy, it's easier to think of the pyramids as representations of either multiple objects or a single object.

The former approach suggests each pyramid is the representation of a different object. Since each object is represented by a similar structure, it's reasonable to assume all of the objects are similar—like three different stars. The size of each pyramid is possibly informing us about the measure of a property of the object it represents that happens to be a property the three objects have in common.

Using the latter approach, we can entertain the idea that the pyramids represent a progression of measures of the same property of a single object. In other words, each pyramid of a different size represents a different state—a different measure—of some property of an object where the states form a progression. (Consider a generic situation with a progression from *uncertainty* as a measure of some property of an object to *certainty* as a measure of the same property of that object. If these two measures are separated by a third measure of the object's property, then you might think a measure consistent with that progression would be *certainty and uncertainty*.) It's hard to resist the temptation to think of a progression as temporal. In a causally connected world where change is observed in a spatiotemporal context, we can't imagine the measure of a property changing without the passage of time.

At this point in the presentation, instead of thinking about the progression in size of the pyramids in terms of the passage of time, think about it in terms of something inherently related to the passage of time, namely the progression from uncertainty to certainty or vice versa. A progression from uncertainty to certainty implies a sense of dynamics, but it's also possible to interpret the pyramids from a more static perspective. What if each pyramid, located at a different measure along the spectrum from uncertainty to certainty, represents a different *mode of observation* of an object? Referring to each pyramid as a mode of observation of an object introduces an interpretation the pyramids can accommodate. (It's not an obvious interpretation.) The question is why might it be advantageous to distinguish those modes? Should three modes of observation of an object imply three modes of appearance of that object? We'll come back to this idea shortly.

The Map

There's a popular theory regarding the pyramids offering a convenient starting point for further reflection. In their book The Orion Mystery, Bauval and Gilbert argue that three stars in Orion's belt would have periodically aligned in the night sky with the pyramids of Giza, an alignment that dates as far back as 10,450 BC. They suggest the pyramids' builders were aware of the alignment. If it's not a coincidence, the alignment implies the pyramids are essentially a representation or a map of something they were designed to bring to our attention. At some point in our lives, all of us employ rocks on the ground, lines in the dirt, or pictures on paper in an attempt to construct a map. In each case the rocks, the lines, and pictures on paper represent features we deem important enough to add to our map to make communication of a situation easier. The maps we're describing have something else in common: each is inherently fragile and unlikely to survive the passage of time. Fragility isn't an issue for the pyramids. There may be uncertainty and debate regarding when the pyramids were built, but there's no doubt they've been there for a very long time desperately trying to tell their story. Perhaps the alignment of pyramids with stars in the constellation Orion is a clue to knowledge that could be of value.

Let's stick with the idea that the design of pyramids encodes some kind of map. Before we can address why the map was created, we need to understand what's being mapped. To help us understand what's being mapped, we need to refer to the map's legend. Where might we find that legend? Given the pyramids are three dimensional, the legend of the map is probably a three dimensional structure that's undeniably a part of the complex but sufficiently separated from the pyramids.

The obvious candidate for a legend is the Sphinx. The structure is sufficiently separated from the pyramids and it's symbolic. The Sphinx as legend of a map explains why there's only one of them where some have suggested there should be two: one Sphinx invites us to contemplate what it symbolizes in a way that two would not. So what is the head of a man on the body of a lion trying to tell us? It's telling us the pyramids are a map of human nature. A map of human nature just might be profound enough, if it's fundamental enough, to warrant the effort required to build the pyramids. If the pyramids are a map of human nature, could it be that searching for the meaning of the pyramids also provides insight into human nature?

The search for meaning in our lives is part of what makes us human. All creatures in nature know what they're supposed to do, all except man. Man is free to choose how he lives his life and how he finds meaning in life. Freedom, however, comes with a price: man feels the tension between what he is and what he can be. He feels the need to fulfill a sense of purpose in the context of his world and its limitations. Regardless of whether humanity is as unique in the universe as we're inclined to believe, tensions that motivate us and the need to resolve them may be symptomatic of a cosmic dynamic more fundamental than what we typically attribute to human nature.

An Interpretation

The above discussion might lead one to interpret the pyramids as representations of three stars in the constellation Orion at the time of their alignment with the pyramids. The problem with this interpretation is obvious: pyramids don't physically resemble stars. In fact, the connection of the pyramids at Giza with stars in Orion is based primarily on the observation that each pyramid's position relative to the others' aligns with three stars in Orion displaying the same relative positioning with respect to each other. Suppose we concede that the pyramids could represent three stars even without a physical resemblance. If this concession is made, the pyramids represent stars, but the downside is they're unable to represent anything else; no other object could be represented as part of the message of the pyramids. The resulting loss of generality isn't consistent with the proposition of a profound message. Achieving the robustness we're looking for requires a different approach.

What if the three stars in the constellation Orion, in the context of their connection with the pyramids, are meant to represent an object in three *modes of appearance*? In other words, the stars in Orion represent the idea that three modes of appearance are associated with an object, any object, where the three modes of appearance are related in a manner suggested by the position of each star relative to the others. If such is the case, then what do the three pyramids at Giza and their alignment with Orion's stars represent? The pyramids represent an object in three *modes of observation* where the three modes of observation are related in a manner suggested by the position of each pyramid relative to the others. Even without additional details regarding the nature of these modes, positional alignment between the stars in Orion's belt and the pyramids of Giza implies the familiar concept of an *object observed as it appears*. Fortunately, the analysis can be simplified going forward by changing focus.

The subtlety of the connection between Orion's stars and the pyramids suggests the alignment between three modes of an object as it appears and three modes of an object as it's observed represents a clue supporting a more concise interpretation of the pyramids independent of the stars in Orion. *What's desired is an additional interpretation of the pyramids, and only the pyramids, reflecting the result of what happens when three modes of an object as it appears are combined with three modes of an object as it's observed to produce three states describing some form of phenomenological apprehension of an object observed as it appears.*[1]

There is the possibility of a more concise interpretation if we introduce what phenomenologists describe as *intentionality*.[2] In a phenomenological situation, intentionality is a network of interactions between the subjective and objective connecting three modes of an object as it appears with three modes of an object as it's observed. As part of the resolution of that network of intentionality, each pyramid represents one state of perception of (an object). To be more precise, each pyramid is the phenomenological expression of an uncertain state, a state both certain and uncertain, or a certain state, such that all three states are represented.

The details of intentionality underlying the phenomenological apprehension of an object are outside of the scope of this presentation. It's a convenient omission because the details are more tedious than anyone except a phenomenologist might care to endure, but it's an unfortunate omission in the sense that you need details—particularly with the way we've defined intentionality—to fully appreciate how subjectivity as perception of (an object) and objectivity as consciousness come together.[3] The network of intentionality is developed in the primary document (Kodiak 2014).

Chapter two will sidestep a discussion of the details of intentionality by abandoning the distinction between three separate modes of an object as it appears and three separate modes of an object as it's observed in favor of a less detailed perspective that preserves the distinction between the object as it appears and the object as it's observed. What will make the simplification possible is a temporal metric described as the *object-observed-as-it-appears,* a metric that positions the object as it appears and the object as it's observed between uncertainty and certainty, respectively. To be more explicit, the object-observed-as-it-appears represents the object as it appears when the state of the metric is sufficiently uncertain, and the object-observed-as-it-appears represents the object as it's observed when the state of the metric is sufficiently certain. The implication is there's an interim state where the object-observed-as-it-appears is neither sufficiently uncertain nor certain. The object-observed-as-it-appears in that particular state can be claimed to represent the object observed as it appears in the *now*. (Adding dashes to the phrase 'object observed as it appears' represents an attempt at distinguishing when the phrase is used to reference an object capable of presenting itself in multiple states as compared to using the phrase without dashes to describe an object in one particular state of interest, namely the state of being observed as it appears.)

This might be a good time to recap our approach. The alignment of the pyramids with three stars in Orion's belt can be interpreted as suggesting three modes of an object as it's observed are interrelated with three modes of the object as it appears. Details of the interrelationship are less than obvious. We've also suggested that the interrelationship is part of what philosophers, linguists, and psychologists refer to as intentionality. Rather than get bogged down in that level of detail, we can limit our perspective to what intentionality enables: the map can be interpreted as representing the phenomenological apprehension of an object as it progresses from an object as it appears to an object as it's observed. Using the pyramids, the subjectivity of the phenomenological apprehension of an object will be presented in chapter two as perception of (an object) when that object is observed as it appears in the now. The implication is the designers of the pyramid complex aligned the pyramids with stars in Orion to give us a clue to a robust conceptual framework capable of accommodating a number of situations.[4] (See footnote 1 in the preface.)

The approach we've recapped fails to mention consciousness. Although consciousness will be discussed briefly in chapter two, chapter three focuses on the relationship between objective consciousness and subjective perception manifesting as consciousness of (perception of (an object)). As you work through that chapter, you might be inclined to speculate on how intentionality enables the relationship between consciousness and perception.

Dimensions of the Situation

In this section we'll find it convenient to designate the three largest pyramids of Giza as primary pyramids to distinguish them from the smaller subsidiary pyramids that are also part of the monument. Let's assume the primary pyramids can be interpreted as the subjective expression of the phenomenological apprehension of an object. What about the subsidiary pyramids at the complex? What might they represent?

A bird's eye view of the pyramids, as shown in figure 1-1, reveals seven subsidiary pyramids: three on the east side of the Great Pyramid, one on the south side of the middle pyramid, and three on the south side of the smallest pyramid. The symmetry of groupings of subsidiary pyramids with respect to the primary pyramids suggests the subsidiary pyramids are there to provide additional information; however, it's not clear whether the subsidiary pyramids are trying to inform us about perception of (an object) or about what's more fundamental than perception of (an object). If the subsidiary pyramids are truly subsidiary to the primary pyramids in more than size, it's reasonable to assume the information they add is contextual in the sense of providing morphological details at a more fundamental level than we can observe.

The difficulty with interpreting what the subsidiary pyramids represent is twofold: First, the generic situation described by the pyramids of Giza—what you see in figure 1-1—allows the primary and subsidiary pyramids to support multiple interpretations. The subsidiary pyramids may be easier to describe with respect to the generic situation, the approach we'll take, but easier to understand with respect to the phenomenological situation. Second, interpretation of the subsidiary pyramids cannot be adequately presented without delving further into the foundations of the frameworks developed in the primary document than we're prepared to explore

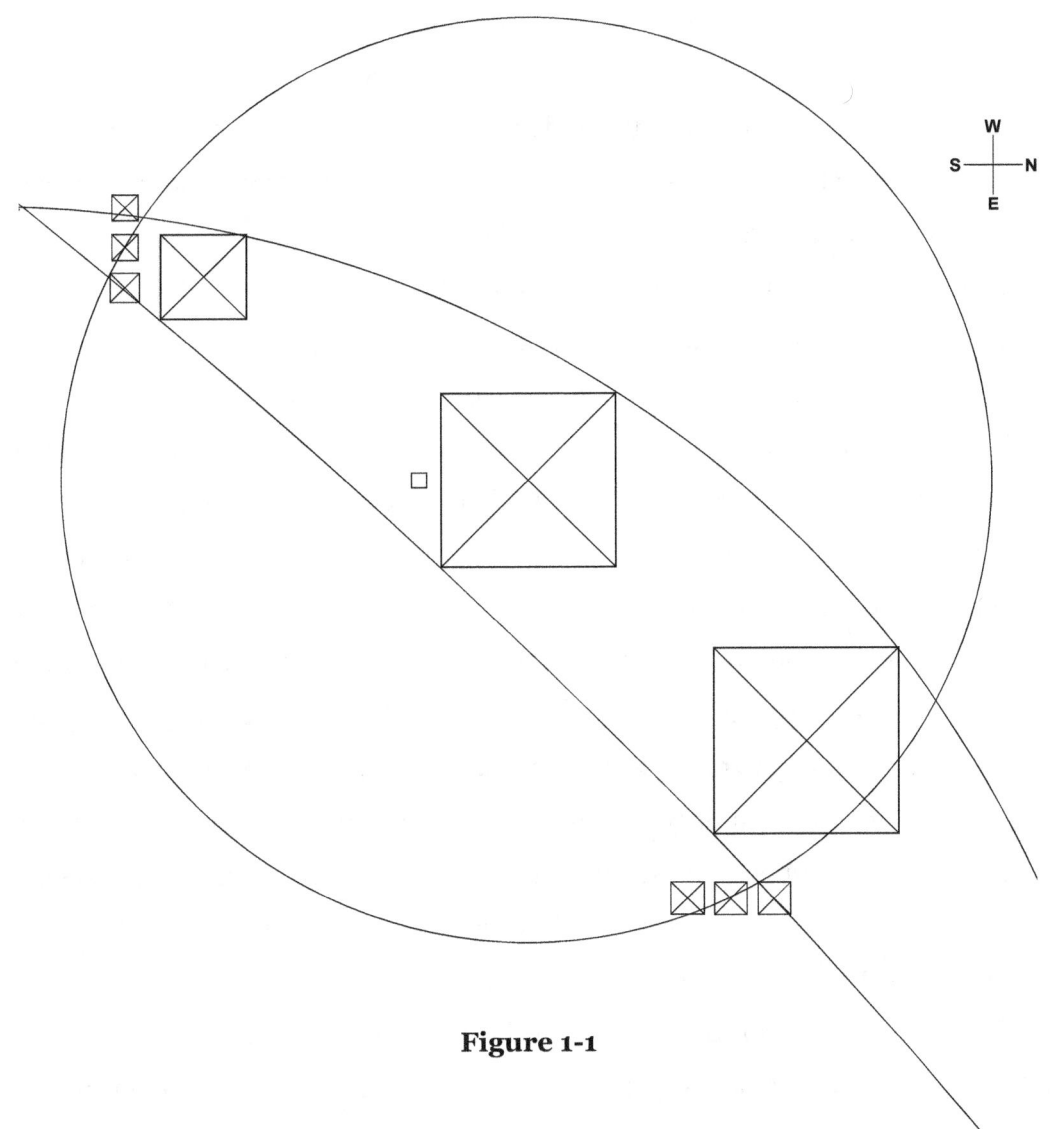

Figure 1-1

The Search For Meaning

in this presentation. So, at the risk of raising more questions than we answer, let's address the subsidiary pyramids by suggesting a curiosity regarding them represents more than a curiosity. (This initial interpretation ignores the lone subsidiary pyramid on the south side of the middle pyramid, due to its smaller size, to focus on the six remaining subsidiary pyramids clustered into two groups of three pyramids each.)

There's a geometric relationship between the six subsidiary pyramids associated with the smallest and largest primary pyramids that's best explained via an illustration. Note the parallel lines in figure 1-2 connecting the centers of the subsidiary pyramids with the corners and center of the middle primary pyramid.[5] When you interpret a figure, you interpret what you see, but sometimes you have to interpret what's suggested by what you see. What's suggested by figure 1-2 is the situation illustrated in figure 1-3 where there's information in the idea that the middle primary pyramid can be replaced by three subsidiary pyramids. The reason three subsidiary pyramids require an interpretation to substitute them for the middle primary pyramid is due to a limitation of the presentation medium: you can't build three additional subsidiary pyramids where you've already built a primary pyramid. You need the middle primary pyramid to communicate the primary message, but you also need three additional subsidiary pyramids, in the same location, to communicate a secondary message supporting the primary message. The scenario in figure 1-3 is a brilliant compromise on the part of the designers that shouldn't surprise anyone who's spent any time exploring the geometry of the structures.

If each of the nine subsidiary pyramids in the figure represents a *fundamental pattern,* the same fundamental pattern, and if each of the two remaining primary pyramids represents that fundamental pattern, then the map at the Giza Plateau might represent a generic situation described by an eleven dimensional interaction of a fundamental pattern.[6] (A detailed discussion of the fundamental pattern each pyramid represents would probably be of interest, but it would take us in a direction we want to avoid in this presentation.)

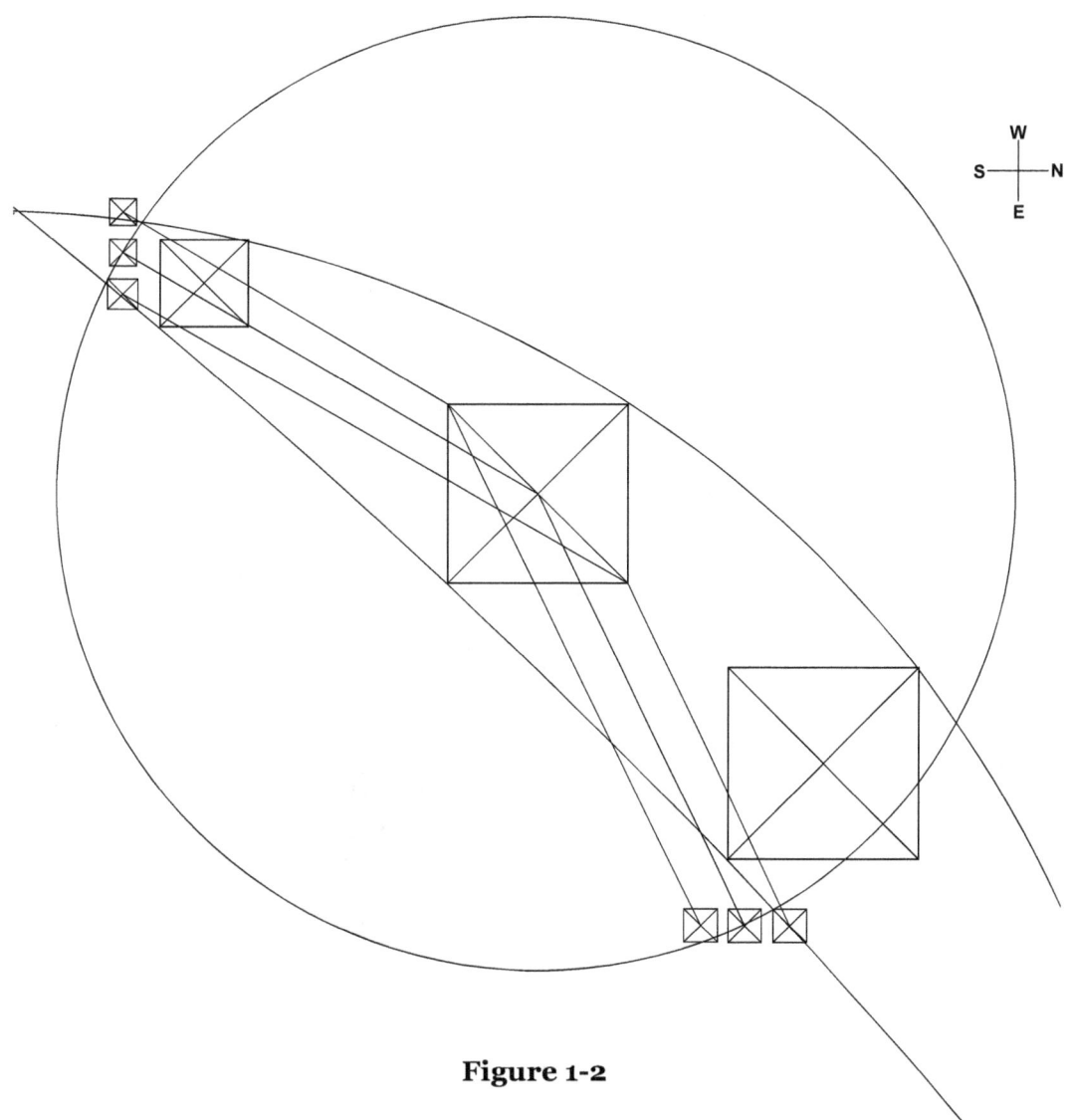

Figure 1-2

The Search For Meaning

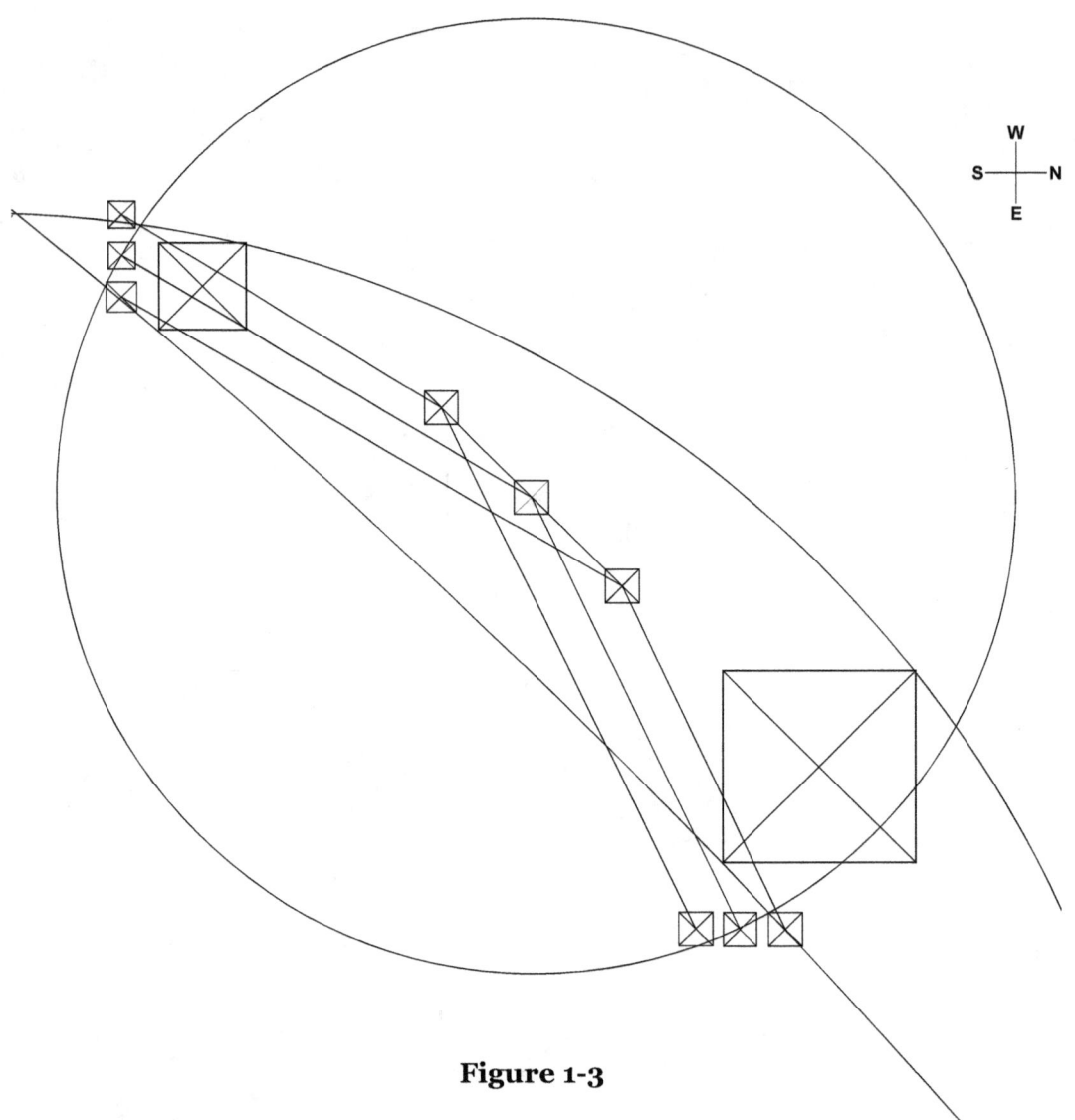

Figure 1-3

Associating a fundamental pattern with dimensionality is a conceptualization the patterns conform to when they organize themselves into a system of interacting patterns. Fortunately, we can avoid that level of detail. All we need for our interpretation is the idea that a network of interacting patterns, however they manifest, coordinate intentionality to interrelate multiple modes of an object as it appears and as it's observed in a manner we understand as consciousness of (perception of (an object)). Although it's not the only interpretation, a generic situation that can be interpreted as a phenomenological situation suggests the message of the pyramids can be understood from the perspective of phenomenology.[7] (Our use of parentheses in describing the situation is reminiscent of a functional relationship. Perception is a function requiring an object, and consciousness is a function requiring perception of an object. Using this shorthand is convenient but it has limitations–particularly with respect to consciousness. To be clear, there's more to the relationship between consciousness and perception than a functional description communicates.[8])

The proposal you've just read with respect to the subsidiary pyramids exposes an observation regarding our approach it's reasonable to suspect was intended by the designers of the Giza complex: the conceptual leap required for interpretation of the subsidiary pyramids is a lot easier to make once you're in the position of reconciling the map presented by the pyramids with an understanding of the conceptual geometry the map's describing. For the most part, we didn't start with an interpretation of the pyramids guiding us to cosmological insights; we started with multidisciplinary insights and subsequently recognized how their underlying conceptual geometry bore a certain familiarity with the design of the Giza complex. The designers of the pyramids were fully aware that understanding their message would be contingent upon a level of understanding of the cosmos. Details of the message the map has to offer were never intended to be intuitively obvious. The designers counted on the gestalt of the monument being intriguing enough to keep us wondering until we understood the connection between the pyramids and the order of things.

There's an additional interpretation of the subsidiary pyramids with the advantage of providing a role for the small subsidiary pyramid to the south of the middle primary pyramid. The smallest and largest primary pyramids both have three subsidiary pyramids. The two sets of three subsidiary pyramids suggest the smallest and largest primary pyramids represent concepts with spatial (three dimensional) interpretations, and the lone subsidiary pyramid suggests the middle primary pyramid represents a concept with a temporal (one dimensional) interpretation. From a phenomenological perspective, the smallest and largest primary pyramids represent the perceptual observation and appearance, respectively, of an object existing in a world–existence in a world we typically comprehend as spatial–and the middle primary pyramid represents an object observed as it appears defining a now we typically comprehend as temporal. You'll see this interpretation in detail in chapter two.

Where We Are and Where We're Headed

Our interpretation suggests a familiar phenomenological situation is built on a network of interacting patterns where each manifestation of a fundamental pattern can be conceptualized as a dimension. If it's possible to interpret the pyramid complex as a map of consciousness of (perception of (an object)), then figure 1-3 is suggesting the phenomenological situation is the interpretation of a nine dimensional generic situation that could be as large as eleven dimensions.

Is the generic situation represented by the pyramids telling us the universe is eleven dimensional? Not necessarily. The generic situation is suggesting the universe–at the very least, some organismic situation in the universe–can be understood as multidimensional, a possibility we're finally catching on to thanks to theoretical advances such as string theory.[9] Here's something to think about: if each dimension each pyramid represents is described by a fundamental pattern with its own dimensionality, then a cosmos enabling this conceptualization is possibly even more multidimensional than what's already hypothesized. The nine or eleven dimensions underlying the phenomenological, psychological, or ontological interpretation of the generic situation might represent an interim step in a necessary reduction in dimensionality required to achieve some goal.[10] Suggesting how a reduction in dimensionality is realized is part of the message of the pyramids.

We started this chapter by proposing the pyramids of Giza represent of the phenomenological apprehension of an object. Now we're suggesting the generic situation represented by the map might reveal information about the multidimensionality of the universe. A detailed discussion of the multidimensionality of the map would provide valuable context, but it's a rabbit hole we probably want to avoid. Our primary goal is to provide enough insight into the map encoded in the design of the pyramids to convince you of the possibility that the map was designed by extraterrestrial visitors to our planet around the beginning of human civilization. Achieving this goal is possible without digressing into cosmological details about the multidimensionality of the universe that aren't well understood and offer little chance of consensus. We do, however, have a secondary goal: introducing research where implications of a multidimensional conceptualization of the cosmos are investigated—research with the potential of providing a better understanding of the message.

Take a quick look at figure 3-14; that's where we're headed. Although we won't develop the argument supporting figure 3-14 with additional details about cosmic multidimensionality, those details are incorporated in the presentation.

Fully deciphering the map presented by the pyramids of Giza requires reconciling its interpretation with our knowledge of the cosmos, knowledge typically expressed via a cosmological model. All cosmological models, because they're models, incorporate some form of implicit or explicit assumption, yet what they all have in common is the need to explain the observable universe. The cosmological model in our research starts with an assumption about the interaction between certainty and uncertainty (Kodiak 2014).

We haven't communicated anything specific regarding the science of cosmology, but perhaps more is communicated than you realize. If physics and cosmology are of interest, then the task at hand is recognizing the cosmological implications of the generic situation from the perspective of physics. Undertaking that task is easier to justify when the generic situation is a multidimensional network with the potential of explaining why the phenomenological situation unfolds in the manner described in the next two chapters.

Although it's beyond the scope of this presentation, a fascinating discovery made with respect to specific situations with an internal observer is the emergence of an organismic sense of time. Since this emergent sense of time is a manifestation of the conceptual geometry, one has to wonder about the underlying physics.

Before moving on, it's time to say goodbye to the subsidiary pyramids. A reconciliation of the design of the pyramids with the phenomenological situation they map can be achieved without further discussion of the subsidiary pyramids.

Chapter 2

Perception of (an Object)

Introduction

We'll begin by briefly discussing the distinction between subjectivity and objectivity in a phenomenological situation described as consciousness of (perception of (an object)). This discussion sets the stage for an interpretation of the pyramids at Giza as a representation of perception of (an object).

Important prerequisites are the three levels of abstraction available for conceptualization and the uncertainties of frame of reference, observation, and measurement realized by the relationship between states defining the second level of abstraction.

The material may seem a bit tedious in a couple of sections, with some forward references to contend with along the way, but perseverance will pay off in chapter three where everything comes together.

Understanding the Situation

Understanding the phenomenological situation requires establishing what we mean by subjectivity and objectivity in the context of the phenomenological apprehension of an object.[1] *Subjectivity with respect to consciousness of (perception of (an object)) makes sense when perception of (an object) is understood as subjectivity. The immediate consequence of this proposition is the suggestion that consciousness must be understood as objectivity.* Associating consciousness with objectivity might seem counter intuitive given the tendency to think of consciousness as inherently egocentric, but keep in mind that consciousness is described as objectivity only as it relates to perception of (an object) described as subjectivity in a situation described as consciousness of (perception of (an object)). (It will turn out that it's possible to associate consciousness with objectivity in the situation under consideration without compromising the sense of egocentricity normally associated with consciousness. Chapter three presents more on the sense of egocentricity associated with consciousness; however, a better understanding of this sense requires what will be described shortly as the third level of abstraction. The third level of abstraction of the phenomenological apprehension of an object is the focus of the primary document.)

In the situation described as consciousness of (perception of (an object)), what's referred to as objectivity obtains when the object as it's observed is the object as it appears and it's perceived as such: once an object is perceived to be observed as it appears, the existence of that object in a world is resolved and subjective apprehension as perception of (an object) is ready for objective expression as consciousness. *The perception of an object, when the object is observed as it appears, collapses to consciousness.* The collapse suggests consciousness is the actualization of potential described as perception of (an object).[2] What's interesting is our familiar subject-object dichotomy isn't destroyed by this collapse; it reemerges in the relationship between subjective and objective measures of consciousness. The challenge will be to define consciousness in a manner that its relationship with perception is consistent with other relationships in which consciousness participates.

Since all that's left as a result of the collapse is consciousness, an argument can be made that once perception of (an object) collapses to consciousness, it no longer makes sense to describe consciousness as consciousness of (perception of (an object)). The collapse of perception of (an object) to consciousness suggests consciousness of (perception of (an object)) describes a *transformation* of subjective perception of (an object) to objective consciousness, where, for reasons that will become clear, consciousness is not described as consciousness of (an object). This transformation brings into question a long standing assumption enshrining consciousness as consciousness of something. Note that although it may be possible to deny consciousness is consciousness of something, it's hard to deny that consciousness resulting from a transformation becomes something.[3] Consciousness becomes objective and subjective measures in relationship to each other for the purpose of completing the phenomenological apprehension of an object.

A counter argument claims the separation of consciousness from perception of (an object) isn't as big of an issue as you might think because the situation requires consciousness to be consciousness of something. Consciousness is understandable as an objective expression of subjective perception of (an object).

Consciousness as objectivity is an interesting proposition you may feel the need to reflect upon at some point, but the more immediate goal of understanding the map presented by the pyramids is best approached by first attempting to understand the three primary pyramids from the perspective of subjectivity described as perception of (an object). Since perception is only half of the story, focusing on it will only get us so far before it becomes necessary to recognize how the map incorporates consciousness.

Levels of Abstraction

Before continuing, it should be emphasized that the discussion will cover three levels of abstraction where one level of abstraction is built upon another. The first level is the most primitive and most general, and the third level is the most complex and most specific. Let's examine each of the three levels in more detail.

The first level of abstraction is represented by a sufficiently primitive measure of uncertainty in a multidimensional space representing uncertainty. We don't specify exactly what the measure of uncertainty is but a reasonable example of such a measure might be along the lines of the concept of entropy or—what's equally cavalier—dimensionality.[4] The downside of this level of abstraction is that in order to discuss a particular context we need more sophisticated abstractions capable of communicating functionality.

The second level of abstraction brings us closer to supporting different contexts in the sense of providing measures of certainty and uncertainty from the perspective of a subject-object dichotomy. Although it's more sophisticated than the primitive first level it's built upon, the second level of abstraction distinguishes only four basic measures defined by subjective and objective measures of certainty and uncertainty. These measures are interrelated in a manner that will turn out to be fundamental to our understanding of the phenomenological situation. Because of the generality of this level of abstraction, the measures can be made more intuitive by referring to them as *states*. It's convenient to refer to these measures as states of objective certainty, subjective certainty, objective uncertainty, or subjective uncertainty. We can also drop the states qualifier when the context is obvious and simply refer to them as objective certainty, subjective certainty, etc. The second level of abstraction will play a significant role in understanding the map, so there's a lot more to be discussed regarding these states and how they're interrelated.

Finally, representing phenomenology, ontology, or psychology requires a third level of abstraction built on the two lower levels. Measures at the third level of abstraction should present a schematization of what's of interest. With respect to phenomenology, every measure describing phenomenology (at the third level of abstraction) is also a state of objective or subjective certainty or a state of objective or subjective uncertainty (at the second level of abstraction) built upon a primitive measure of uncertainty (at the first level of abstraction).

(We're not suggesting all measures of phenomenology at the third level of abstraction representing states of objective uncertainty at the second level of abstraction, for example, have the same underlying value of uncertainty at the most primitive first level of abstraction. The value of uncertainty, measured at the first level of abstraction, of a state at the second level of abstraction is determined by the distance of that state from its objective frame of reference—an objective frame of reference that represents a state of objective certainty. Frames of reference will be discussed in more detail shortly. At the moment, it's sufficient to note that caution is required when describing measurements of uncertainty from the perspective of multiple objective frames of reference in a space with one certainty frame of reference. Fortunately, measurements of uncertainty of interest to us will turn out to be measurements of uncertainty with respect to a certainty frame of reference.)

To keep the presentation as general as possible, the plan is to limit details of the discussion to a description of measures at the first two levels of abstraction. Implementing this plan requires hedging our bets because the lower levels of abstraction can make it challenging to appreciate implications of the map. So, to facilitate a better understanding of what's presented, the map will be interpreted within the context of phenomenology described as consciousness of (perception of (an object)), but the interpretation will be below the full level of detail of what you might expect of a conceptual framework. (Don't lose sight of the idea that the map is applicable to more than the phenomenological situation used to illustrate its message.)

As the presentation progresses, we'll continue to reference conceptual frameworks of the primary document and reference a more fundamental foundation manifesting a conceptual geometry. It should become clear that the conceptual frameworks articulate the more specific third level of abstraction, and the conceptual geometry illustrates how the more general first and second levels of abstraction come together. These three levels of abstraction are more familiar than you realize. Think about the mind (the third level of abstraction) as it relates to the brain (the second level of abstraction) and the brain as it relates to the body. The analogy would be obvious if the presentation offered more details regarding homeostasis at the first level of abstraction or more about plasticity at the second level of abstraction.

Where the Pyramids Fit In

Interpreting the pyramids as a representation of the subjective apprehension of an object requires forgoing the simplicity of describing each pyramid as a measure of the same property of an object—an idea suggested in chapter one—because that approach fails to emphasize the distinction between perception and the object perceived inherent to perception of (an object). The good news is we can simplify things quite a bit by making use of the object-observed-as-it appears metric introduced in chapter one. With respect to the subjective apprehension of an object, regardless of whether we're discussing perception or the object perceived, the distinction of interest will be how the object as it's observed relates to the object as it appears.

Let's propose a progression from uncertainty to certainty can be represented by a progressive reduction of some measure of uncertainty in a multidimensional space representing uncertainty at the first level of abstraction. Assuming the three pyramids can be situated within such a space, is there anything about them providing an idea of the relative value of each pyramid's uncertainty? If the progression in size of the pyramids provides a clue to their uncertainty in a space representing uncertainty, then—in the context of what will turn out to be an expression of the subjectivity of a situation described as consciousness of (perception of (an object)) where subjectivity is realized as perception of (an object) when that object is observed as it appears—we can associate the largest pyramid (the Great Pyramid) with a state of uncertainty, the smallest pyramid with a state of certainty, and the middle pyramid with a state both certain and uncertain. (There's no claim that the size of each pyramid represents its measure of uncertainty. What we're suggesting is a correlation between the two.)

Needed sophistication ensues if the middle pyramid is associated with something conceivable as certain and uncertain we understand as the temporal now and assumptions are made regarding the relationship between the now and what can be interpreted as a progression from uncertainty to certainty represented by the pyramids. Let's assume the largest pyramid, representing a state of uncertainty, can be associated with before the now, and the smallest pyramid, representing a state of certainty, can be associated with after the now. These associations and assumptions suggest the pyramids are capable of representing a temporal progression where before the now is inherently more uncertain than after the now. This is a good start, but we still need additional insight to relate a temporal progression represented by the pyramids to perception of (an object).

A temporal direction is implied when a progression from uncertainty to certainty is correlated with a progression from before the now to after the now. Since an object appearing in the future can be argued to be unknown and only becomes known by progressing from uncertainty of the object as it appears in the uncertain future to certainty of the object as it's observed in the certain past, it's likely that perception of (an object) accommodates a progression from the *object as it appears as perception* before the now to the *object as it's observed as perception* after the now. The object as it appears as perception needn't be the same as the object as it appears, and the object observed as perception needn't be the same as the object as it's observed. It's reasonable to suspect we're interested in a temporal progression from an object as it appears, to the object as it appears as perception, to the object as it's observed as perception, to the object as it's observed.

Before we offer a proposal regarding the representation of perception by the pyramids, let's be clear about two intuitive assumptions: the nature of perception is such that an object can only be successfully observed as perception if it appears as perception, and perception after the now is more certain than perception before the now. *What becomes important is if an object appears as perception before the now and is observed as perception after the now, then the temporal implication is perception of (an object) describes the perception of an object observed as it appears in the now.* Using the suggestion of the previous paragraph, one can also infer there's an object as it appears before the now and before the object's appearance as perception, which means it's more uncertain than the object's appearance as perception, and there's an object as it's observed after the now and after the object's observation as perception, which means it's more certain than the object's observation as perception.[5] Keep these assumptions in mind when we discuss figures 2-1 through 2-3.

The object as it appears as perception can be represented by the largest pyramid, and the object as it's observed as perception can be represented by the smallest pyramid. These two pyramids can be assumed to represent perception. What's left is the object observed as it appears in the now represented by the middle pyramid.

We're talking about perception of (an object) as a progression from an object as it appears as perception to an object as it's observed as perception, a progression of perception contingent upon the temporal progression from uncertainty of an object as it appears to certainty of that object as it's observed. The object, represented by the object-observed-as-it-appears, is observed as it appears in a now which is neither completely certain nor uncertain. As we've already mentioned, the representation of perception of (an object) by the pyramids is an expression of the subjectivity of a situation that's understandable as consciousness of (perception of (an object)). Consciousness in this context can be described as the actualization of potential represented by perception of (an object) when the object perceived is observed as it appears in the now—an object that must exist in a world. To be more concise, perception of (an object) collapses to consciousness. There is, as you might have guessed, an obvious question that comes to mind: where in the pyramid complex is the representation of consciousness? Consciousness will be introduced in chapter three.

Now that what's expected of the pyramids in terms of representing perception of (an object) is beginning to unfold, this would be a good time to introduce additional details essential to understanding how states at the second level of abstraction relate to each other.

The Progression of States

The world we live in requires distinctions between what's certain and what's uncertain. Making those distinctions requires separating the state of certainty into states of objective and subjective certainty and separating the state of uncertainty into states of objective and subjective uncertainty. As mentioned earlier, these four states occupy the second level of abstraction in an order of things comprehendible via a subject-object dichotomy.

Since values of uncertainty of states of certainty are smaller than values of uncertainty of states of uncertainty, regardless of whether the states are subjective or objective, we need to make an additional assumption that objectivity is more certain than subjectivity, except the state of subjective certainty is more certain than the state of objective uncertainty. *We're now in a position to propose that order from uncertainty to certainty is implied by a progressive reduction in uncertainty from a state of subjective uncertainty, to objective uncertainty, followed by subjective certainty, and finally objective certainty.* This progression from states of uncertainty to states of certainty will be illustrated in the subject-object dichotomies of scenarios in figures 2-1 through 2-4, and it also happens to be the progression from states of uncertainty to states of certainty of the subject-object dichotomies supporting the phenomenological situation we'll encounter in the next chapter.

The progression of states offers a clue to how states at the second level of abstraction relate to each other beyond the obvious distinction between subjectivity and objectivity, raising the question of whether there's anything else about the relationship between the states that might be of value to our understanding of them or our understanding of how they can be utilized? The next step toward understanding interrelationships between the states—let's call it their *morphology*—requires a discussion regarding frames of reference.

Frames of Reference

Progressive reductions in uncertainty reveal a convenient frame of reference when uncertainty measured at the first level of abstraction is reduced to zero. At that point, there's certainty defining what can be described as a *certainty frame of reference*. (As we'll discover in figure 2-4, the map encoded in the pyramids is capable of accommodating subjective perception or objective consciousness, yet perception and consciousness won't share the same certainty frame of reference.) Although we can identify scenarios where an objective frame of reference represents a certainty frame of reference, an objective frame of reference isn't limited to occurring at a certainty frame of reference. An objective frame of reference is so named because it stands in contrast to a subjective frame of reference. Objective and subjective frames of reference are essentially technical terms deriving their meaning from how we choose to use them.

If the value of uncertainty of a state of objective certainty is smaller than the value of uncertainty of a state of objective uncertainty, then the likely candidate for an *objective frame of reference* in any particular scenario is whatever represents a state of objective certainty. Similarly, if the value of uncertainty of a state of subjective certainty is smaller than the value of uncertainty of a state of subjective uncertainty, then the likely candidate for a *subjective frame of reference* in any particular scenario is whatever represents a state of subjective certainty.

The Uncertainty of Observation

Since the states of objective and subjective certainty qualify as frames of reference, the difference between measures of uncertainty of states of objective and subjective certainty—think of it as a distance—represents what can be described as *uncertainty of the frame of reference*. We can build on this description if the measure of uncertainty represented by the distance from the state of objective certainty to the state of objective uncertainty represents what can be described as the *uncertainty of observation,* and the measure of uncertainty represented by the distance from the state of objective certainty to the state of subjective uncertainty—what will always be the largest measure of uncertainty—represents what can be described as the *uncertainty of measurement.*

The uncertainties are related such that uncertainty of the frame of reference of an observation plus uncertainty of measurement of the observation equals the uncertainty of observation. For this relationship to be true, all uncertainties are measured with respect to an objective frame of reference represented by the state of objective certainty, and uncertainty of the frame of reference reveals itself—in the progression of states of interest to us—as a negative adjustment to the uncertainty of measurement.[6] (If instead of treating the uncertainty of measurement as uncertainty measured from the perspective of subjective certainty, we treat it as measured from objective certainty, it doesn't eliminate uncertainty of the frame of reference. Uncertainty of the frame of reference emerges as an uncertainty of range.)

What's interesting about the uncertainties of observation, frame of reference, and measurement is conceptually they seem intuitive when thought of in the context of perception of (an object). These uncertainties may be intuitive with respect to perception, but they're arguably more fundamental than one context: the underlying states of objective and subjective certainty and the states of objective and subjective uncertainty, at the second level of abstraction, are capable of supporting situations at the third level of abstraction that aren't limited to a phenomenological interpretation. The implication is uncertainties of frame of reference, measurement, and observation made possible by states at the second level of abstraction needn't be limited to supporting perception of (an object).

The next section will return to our discussion regarding the representation of perception of (an object) by the pyramids. That section will start by parsing details of the object-observed-as-it-appears as a progression from uncertainty to certainty to put insight provided by the pyramids into perspective with respect to perception. It won't be the last word regarding the map provided by the pyramids, but, at this level of detail, it's as far as we can to take the interpretation before considering the entire situation described as consciousness of (perception of (an object)).

Back to the Pyramids

Discussion regarding a frame of reference typically raises the question of whether the frame of reference is egocentric or allocentric?[7] With respect to perception of (an object), the question can be posed as a choice between a subjective object-centric or a subjective perception-centric frame of reference taken from the perspective of either the object-observed-as-it-appears or perception, respectively. With diagrams introduced in this section, we'll argue perception of (an object) when that object is observed as it appears, interpreted from the perspective of a progression from uncertainty to certainty, can't be accommodated without both frames of reference because the metric we're calling the object-observed-as-it-appears starts as subjective with respect to its perception and ends as objective with respect to its perception:

The object-observed-as-it-appears starts as its own object-centric perspective that's also a subjective perspective; it offers a subjective frame of reference of the object as it appears in a subject-object relationship with objective perception. As a result of its progression from uncertainty to certainty, the object-observed-as-it-appears ends as objective in the sense of being the object as it's observed in a subject-object relationship with subjective perception; the subjective frame of reference becomes that of a perception-centric perspective. To state the above more concisely, perception of (an object) transitions from an object-centric perspective to a perception-centric perspective when the object transitions from appearing (inherently object-centric) to being observed (inherently perception-centric). The transition of the object-observed-as-it-appears from appearing to being observed is also a transition from uncertainty to certainty.

They say a picture's worth a thousand words; so, just to be on the safe side, a series of three of them will help get the idea across. Let's discuss the preceding paragraph with the aid of these diagrams.

In figure 2-1, the object-observed-as-it-appears starts as its own object-centric perspective that's also a subjective perspective. This object-centric perspective is the object as it appears represented by the states of subjective certainty (C(s)) and subjective uncertainty (U(s)). The state of subjective certainty associated with the middle pyramid, which also happens to represent the now, represents a subjective frame of reference of the object as it appears; and the state of subjective uncertainty, associated with the question mark on the right, will be speculated in a later section as representing the object as it appears as expectation. In an order of things described by a subject-object dichotomy, an object-centric perspective of the object as it appears represents the subjective side of a relationship with respect to objective perception represented by the states of objective certainty (C(o)) and objective uncertainty (U(o)). C(o) represents the object as it's observed as perception and U(o) represents the object as it appears as perception. In this scenario, these two states are associated with the smallest and largest pyramids representing perception.

Figure 2-2 illustrates what happens when the progression from uncertainty to certainty results in the object-observed-as-it-appears becoming objective in the sense of being the object as it's observed represented by the states of objective certainty and objective uncertainty with respect to subjective perception represented by the states of subjective certainty and subjective uncertainty. Note that the object as it's observed represented by the state of objective uncertainty is associated with the middle pyramid and consequently with the now. In a later section, the state of objective certainty associated with the question mark will be speculated as representing

Figure 2-1

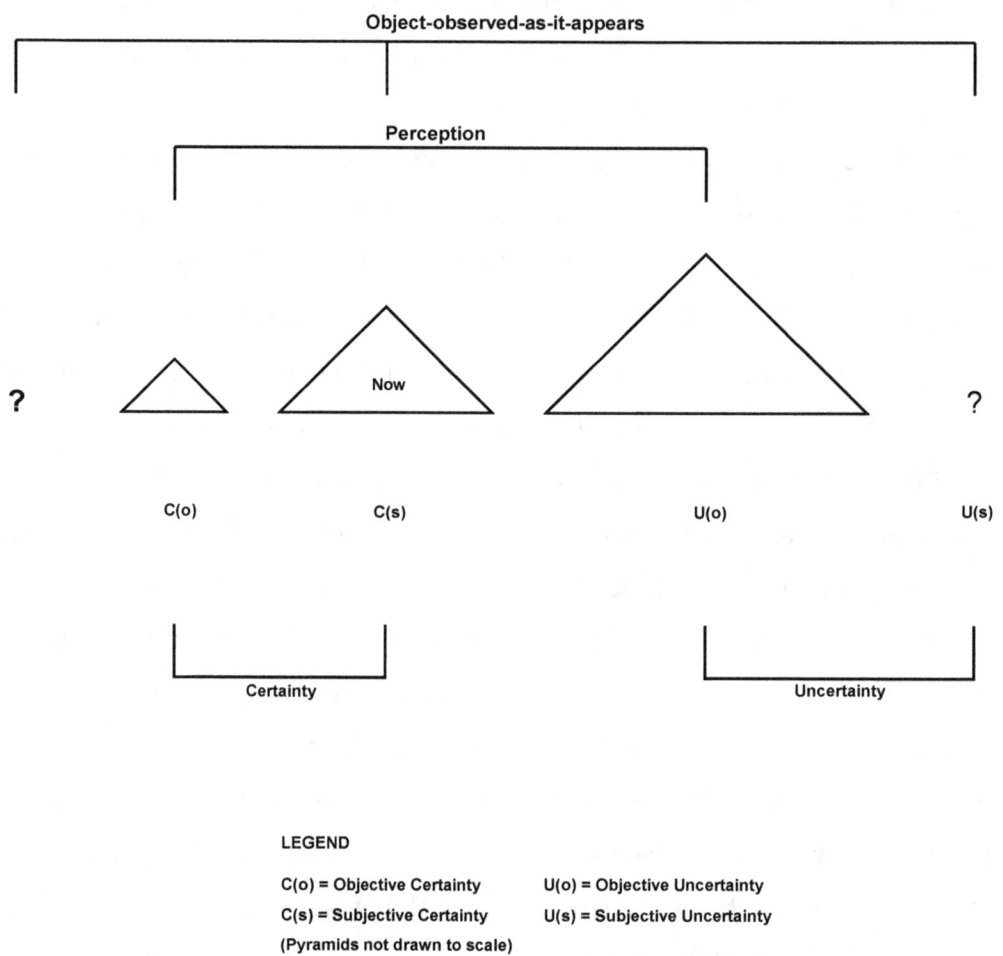

30 According to the Map

Figure 2-2

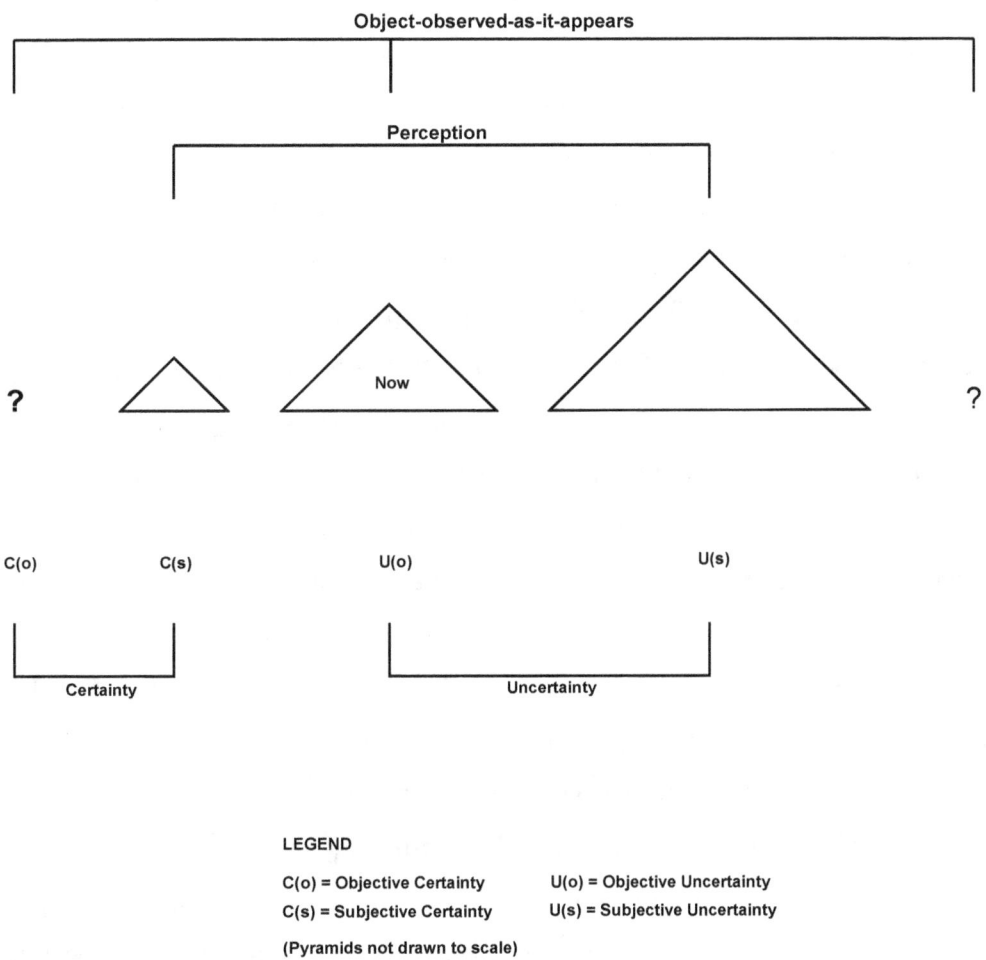

Perception of (an Object)

the object as it's observed as memory. Subjective perception, represented in this scenario by the states of subjective certainty and subjective uncertainty, becomes associated with the smallest and largest pyramids representing perception: the state of subjective certainty represents the object as it's observed as perception, and the state of subjective uncertainty represents the object as it appears as perception. Even though it's not associated with the now, the state of subjective certainty still represents the subjective frame of reference.

(Objective perception in figure 2-1 can't be just the object as it's observed as perception; it must be the object as it's observed as perception and the object as it appears as perception. Similarly, subjective perception in figure 2-2 can't be just the object as it appears as perception; it too must be the object as it's observed as perception and the object as it appears as perception. The reasoning is straightforward: There's one perception from two different perspectives, objective and subjective, and each perspective has the same two components in common, namely the object as it's observed as perception and the object as it appears as perception. Both components are required for an object to be perceived as existing in a world, and both perspectives are required to claim perception of (an object), an object represented by the object-observed-as-it-appears, is perception of an object observed as it appears in the now.)

If you think about the progression from uncertainty to certainty represented by combining both figures, as illustrated in figure 2-3, the object-observed-as-it-appears transitions from a state of uncertainty represented by the state of subjective uncertainty in figure 2-1 (a state representing the object as it appears as expectation) to a state of certainty represented by the state of objective certainty in figure 2-2 (a state representing the object as it's observed as memory).[8] The focal point of the transition occurs in the now. The object as it appears in the now is represented by the state of subjective certainty, and the object as it's observed in the now is represented by the state of objective uncertainty. The temporal coincidence of these states is such that the object can truly be described as observed as it appears in the now.

Figure 2-3

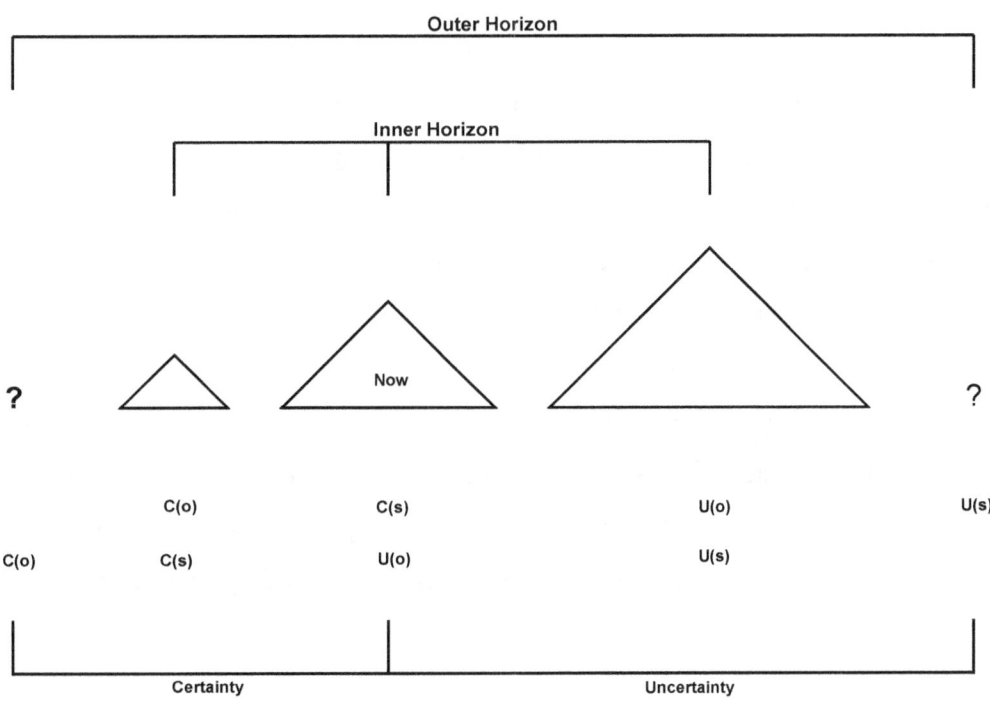

Perception of (an Object)

If we attend to the situation illustrated in figure 2-3 and focus on the four states representing each of the two scenarios the figure brings together, all of the states of certainty occur after the now or in the now, and all of the states of uncertainty occur before the now or in the now. Figure 2-3 offers more specificity with respect to the three states, namely uncertainty, certainty and uncertainty, and certainty, associated with the *pyramids'* representation of perception of (an object) when the object perceived can be claimed to be observed as it appears. The state of certainty (represented by the smallest pyramid) will be either a state of objective or subjective certainty, and the state of uncertainty (represented by the largest pyramid) will be either a state of objective or subjective uncertainty, yet it's still the state transitioning between the certain and uncertain (represented by the middle pyramid) that commands our attention, particularly since it's associated with the now. The transition in the now is a transition of the object-observed-as-it-appears describable as both certain and uncertain–transition from a state of subjective certainty (representing the object as it appears) to a state of objective uncertainty (representing the object as it's observed).

(What's interesting in this interpretation is objective perception in figure 2-1, represented by the states of objective certainty and objective uncertainty, can also be interpreted as a proxy for the object as it's observed. A proxy for the object as it's observed maintains the object-observed-as-it-appears metric associated with the object as it appears represented by the states of subjective certainty and subjective uncertainty on the subjective side of a subject-object dichotomy. Also note that subjective perception in figure 2-2, represented by the states of subjective certainty and subjective uncertainty, can be interpreted as a proxy for the object as it appears. A proxy for the object as it appears maintains the object-observed-as-it-appears metric associated with the object as it's observed represented by the states of objective certainty and objective uncertainty on the objective side of a subject-object dichotomy. If it's possible to interpret perception as a proxy for the object-observed-as-it-appears, then perception of (an object) as a process that progresses from the scenario in figure 2-1 to the scenario in figure 2-2 suggests the object-observed-as-it-appears could conceivably be interpreted as observed *before* it appears. Fortunately, figure 2-3 restrains the temporality of perception to force a sense of simultaneity.)

What's Missing

The temporal interpretation of subjectivity as perception of (an object) is described above as a progression from an object-centric perspective to a perception-centric perspective. Since subjectivity as perception is part of a broader context introduced as the situation described as consciousness of (perception of (an object)), you should expect one more subjective frame of reference associated with consciousness awaits discovery. Although that discovery isn't a journey we plan to undertake here, consciousness probably requires a subjective frame of reference representing the culmination of a progression from an object-centric, to a perception-centric, to an intuition-centric perspective.

The Inner and Outer Horizons

Perception is represented in each scenario illustrated in figures 2-1 through 2-3 by the smallest and largest pyramids, but it's the combined scenario, illustrated by figure 2-3, that's of interest. Before we consider the combined scenario, let's think about the object-observed-as-it-appears: In the first scenario (figure 2-1) the contribution of the outer state to the object-observed-as-it-appears is the state of subjective uncertainty, a measure described as the object as it appears as expectation; and in the second scenario (figure 2-2) the contribution of the outer state to the object-observed-as-it-appears is the state of objective certainty, a measure described as the object as it's observed as memory. (Expectation and memory are terminology to facilitate our comprehension.) What do these measures have in common other than being described as outer states of the object-observed-as-it-appears? In a later section they will be revealed as transcendent measures of the object-observed-as-it-appears.

In figure 2-3, we describe what's represented by all three pyramids as the *inner horizon* and describe the contribution of the outer states to the object-observed-as-it-appears as the *outer horizon*.[9] The measure of the object-observed-as-it-appears represented by the middle pyramid can be described as the object observed as it appears in the now, such that the three pyramids of the inner horizon represent what's best described as perception of (an object) when that object is observed as it appears in the now. This figure suggests the object observed as it appears in the now wouldn't be possible without the progression of the object-observed-as-it-appears from the object as it appears as expectation to the object as it's observed as memory.

The Certainty Frame of Reference of Perception

The most certain state in figure 2-3 appears to be the state of objective certainty associated with the question mark at the far left of the figure, a state representing the object as it's observed as memory. When the object is the object-observed-as-it-appears, as shown in the combined scenario in figure 2-3, the most certain state of objective certainty arguably represents a certainty frame of reference with respect to the entire progression from the most uncertain state to the most certain. The progression of pyramids from uncertainty to certainty in figure 2-3 suggests the same certainty frame of reference, illustrated by the question mark at the far left of the figure, represents the vanishing point of the pyramids. (The vanishing point of the pyramids will be discussed at the beginning of the next chapter. If the argument is unfamiliar, you might benefit from skipping ahead and reading that section before continuing.) *The implication is the certainty frame of reference at the vanishing point represents a measure of zero uncertainty, at the first level of abstraction, from which uncertainty associated with each pyramid can be represented as a distance from the vanishing point to some point representing the amount of uncertainty associated with that pyramid. The vanishing point represents a certainty frame of reference of perception of (an object) when that object is the object-observed-as-it-appears spanning the inner and outer horizons.* (As shown in figure 2-3, distance from the certainty frame of reference is suggested by the horizontal positioning of states; no scale is intended.)

(One needs to be careful: the state of objective certainty in the second scenario of figure 2-3, an objective frame of reference associated with the question mark at the vanishing point of the pyramids, represents what's interpreted as a certainty frame of reference for the entire progression, but the other state of objective certainty, the objective frame of reference associated with the smallest pyramid in the first scenario, doesn't represent a certainty frame of reference. The distance from the certainty frame of reference to a point representing uncertainty of the middle pyramid, the now, isn't affected by our claim that the now represents a state of objective uncertainty in one scenario and a state of subjective certainty in the other scenario, yet we can't claim that uncertainty measured as a distance to a state in an observed-as-it-appears metric can be measured independent of its objective frame of reference. The uncertainty associated with the state of subjective certainty in the now can be measured–if you ignore states associated with perception–from the objective frame of reference in the second scenario, which happens to be the certainty frame of reference, or the uncertainty can be measured from the objective frame of reference in the first scenario. The two measurements of uncertainty will not be equal because each scenario is a progression of states presenting its own observed-as-it-appears metric. The objective frame of reference associated with each state in the now is represented by the state of objective certainty within the progression of states in which the state associated with the now belongs–one progression illustrated by figure 2-1 and the other illustrated by figure 2-2. Uncertainties associated with states of any observed-as-it-appears metric under consideration should all be measured from the same objective frame of reference.[10])

So where is all of this talk about a certainty frame of reference of perception of (an object) leading? The next chapter will argue that consciousness as actualized perception of (an object) can be represented as a *progression of states of consciousness.* The state of objective certainty of that progression, what will turn out to be the apex of the cone of actualization of consciousness, represents an objective frame of reference we can also treat as a certainty frame of reference; however, if consciousness only actualizes potential defined by perception of (an object) when that object is the object-observed-as-it-appears as it's observed as it appears *in the now,* implying consciousness only encompasses the inner horizon of figure 2-3, then how does a certainty frame of reference of consciousness relate to a certainty frame of reference of perception of (an object) where the object is the entirety of the object-observed-as-it-appears spanning the inner and outer horizons? The object-observed-as-it-appears represents an object with a certainty frame of reference defined by the object as it's observed as memory, part of the outer horizon encompassing consciousness. Essentially, what's suggested is the certainty frame of reference of consciousness will not be the certainty frame of reference associated with perception of (an object) identified in figure 2-3. If our illustration is realistic, some form of accommodation of two subject-object dichotomies, one subjective and the other objective, is required. The choice of which certainty frame of reference is relevant depends on whether we're describing subjectivity as perception of (an object) or objectivity as consciousness. Combining the descriptions might require one of the frames of reference to be chosen as the certainty frame of reference. Details regarding these frames of reference are illustrated in the next section.

Conceptualizing Consciousness

If figure 2-3 could be overlaid with the objectivity of consciousness, where would consciousness have to end up to justify the claim that it encompasses the inner horizon defining perception of (an object) but does not encompass the outer horizon defining what will be described in the next section as transcendent measures of the object-observed-as-it-appears? As shown in figure 2-4, consciousness is conceptualized as points A, B, C, and D in a manner that the inner horizon is encompassed by points A and D. The same four points, conceptualizing consciousness, can be claimed to be encompassed by the outer horizon of the object-observed-as-it-appears.[11] When we get to chapter three, you may want to reconcile consciousness conceptualized by points A, B, C, and D in figure 2-4 with the same points representing the actualization of consciousness in figure 3-6. In chapter three, the certainty frame of reference of consciousness at point A is the apex of the cone of actualization of consciousness. If the certainty frame of reference of perception of (an object) is represented by the question mark at the far left of figure 2-4, then what's implied is the certainty frame of reference of perception of (an object) will not be represented along the axis of the cone of actualization of consciousness in any of the figures in chapter three because consciousness doesn't encompass the outer horizon of perception of (an object). (Chapter three will present a caveat—a sidelight to be exact.)

Consciousness is a conceptualization in figure 2-4 because this figure, based upon figure 2-3, focuses on subjectivity manifesting as perception of (an object). Presenting points A, B, C, and D as a conceptualization in figure 2-4 is necessary because consciousness and perception of (an object) are on opposite sides of a subject-object dichotomy where the former is interpreted as objective actualization of subjective potential represented by the latter. In order for consciousness to be an equivalent representation of perception of (an object) consistent with our interpretation of them as causally connected, some set of conditions, a transformation, or an event is assumed to establish equivalence. The event signifying a transformation is described as the collapse of potential represented by perception of (an object) to its actualization as consciousness. Although a direction from perception to consciousness is implied, we're not claiming consciousness simply emerges from perception. The two are intimately related by a network of intentionality.

Figure 2-4

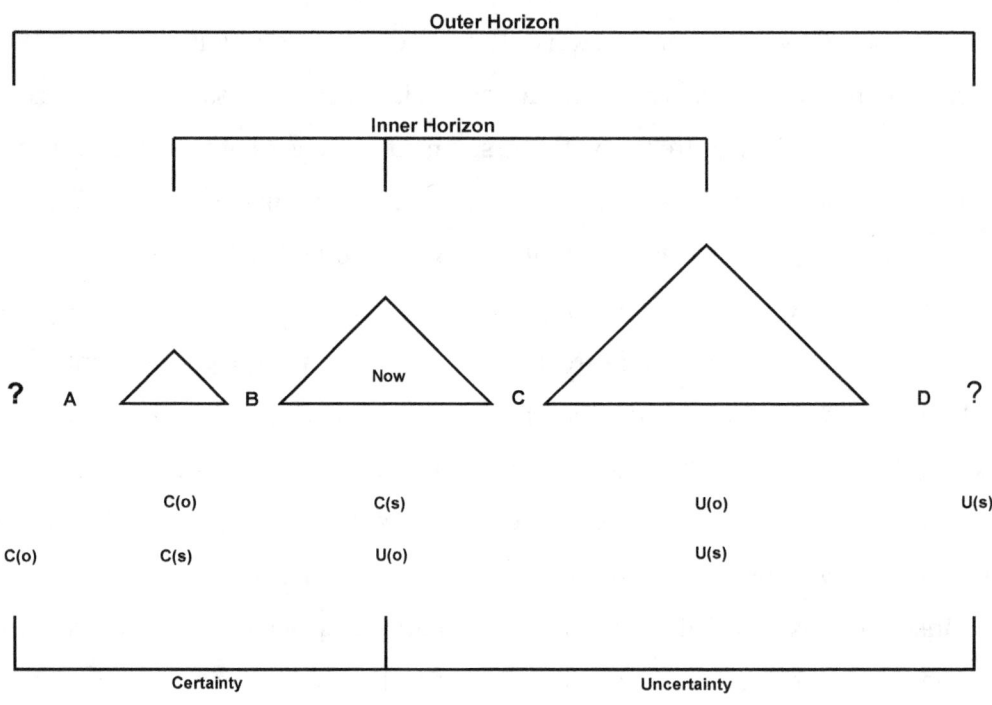

Expectation and Memory

Enough background has been introduced to claim perception of (an object) is intimately related to how the object-observed-as-it-appears changes from its initial state of subjective uncertainty, the object as it appears as expectation, to its final state of objective certainty, the object as it's observed as memory. It's a change in state that reveals a temporal progression, from uncertainty of an expectation to certainty of a memory, consistent with distinguishing the object before the now from the object after the now.[12] In other words, the object as it appears as expectation and the object as it's observed as memory are measures that introduce a temporal context to an object observed as it appears in the now. A measure of the object-observed-as-it-appears described as an expectation or a memory intuitively accounts for the object-observed-as-it-appears beyond the field of perception.[13] Why not focus on the object-observed-as-it-appears as the object observed as it appears in the now? Is the now an expression of time or space? Could it be both? Appreciating why all measures of the metric are important requires introducing another piece of the puzzle.

The object as it appears as expectation and the object as it's observed as memory are *transcendent* with respect to perception of (an object). Perception of (an object) when that object is observed as it appears in the now, defining the inner horizon of figure 2-3, doesn't encompass the outer horizon. In other words, the outer horizon transcends it. You might think it follows from this suggestion that what's *immanent* with respect to perception of (an object) is the object observed as it appears in the now, an object encompassed by perception; but if the object observed as it appears in the now can be assumed to represent something closer to what the object-observed-as-it-appears means, what's more likely is the object observed as it appears in the now is *transcendental* with respect to perception of (an object).[14] If this is the case, then it follows that what's immanent with respect to perception of (an object) is the object as it appears as perception, what's given, and the object as it's observed as perception, what's taken. For the object to appear and be observed as perception, that object must exist in a world that's more than an aggregation of entities; it must exist in a world better described as a system of relations. An object can't appear as perception unless its appearance is in a world, and an object can't be observed as perception unless what's observed is some form of existence. In the way we're presenting it, existence in a world isn't a requirement of the object as it appears as expectation or as it's observed as memory–what we're calling the outer horizon; yet what gives the outer horizon meaning is when it can be described as expectation and memory of an object observed as it appears in the now–a transcendental object whose existence in a world is evidenced by immanent appearance and observation as perception.[15]

(We can describe an object using the concepts of appearance and observation because we haven't limited the concepts to a description of perception. There's the object observed as it appears in the now, transcendental to perception and consciousness, and there's the object as it appears before the now and as it's observed after the now, transcendent to perception and consciousness. Why is a temporal perspective required? A temporal perspective is required if we're trying to make sense of an object as a progression from uncertainty to certainty. The object-observed-as-it-appears is a metric that helps us understand how a transcendental object can be defined by what it's not–neither completely certain nor completely uncertain.)

So the temporal progression of an object–a progression from the object as it appears as expectation, through what we understand as the object observed as it appears in the now, to the object as it's observed as memory–is presented using a metric described as the object-observed-as-it-appears. It's not obvious but another progression may need to be accommodated: the progression from the object-observed-as-it-appears as a transcendent object, as it appears as expectation and as it's observed as memory, to the object-observed-as-it-appears as a transcendental object, observed as it appears in the now. If the former progression represents a temporal progression from before the now to after the now, then what kind of progression should the latter, a progression from the transcendent to the transcendental, represent? The latter may not be a progression; it may represent what's better described as a *focusing* of the transcendent to the transcendental where immanent perception and consciousness represent lenses of the focusing.

What's missing in the description of the focusing is a proposal suggesting how the focusing of a transcendent object to a transcendental object enables that transcendent object to become embedded in an immanent field of perception. The embedding of a transcendent object in a field of perception offers our phenomenological interpretation the possibility of avoiding representationalism. The dynamics of the focusing and how the embedding happens is revealed in the primary document.

(You might think the entire interpretation is sure to be shipwrecked against the rocks of the second law of thermodynamics requiring isolated systems to tend, over time, toward a state of maximum entropy. A progression from an object as it appears as expectation to an object as it's observed as memory—a progression from uncertainty to certainty—appears inconsistent with that law. It's not surprising. The enigma of life is its ability to defy the second law of thermodynamics (Schrödinger 1967). According to the second law of thermodynamics, systems don't organize themselves without the input of organizing information. Is the multidimensional conceptualization suggested in chapter one nothing more than an abstract expression of organismic dynamics, or does it represent a conceptual geometry necessary for the exploitation of information essential to the existence of what we understand as life? In any event, it's a good bet the situation represented by the pyramids already exemplifies the organismic struggle to hold off the second law for as long as possible. In chapter three, the arrow of entropic time is suggested by the progression from certainty to uncertainty of the five pyramids in figure 3-14. For more on thermodynamics and life, see (Schneider and Sagan 2005), and for a foray into time and the second law of thermodynamics, see (Muller 2016).)

Perception as a Process

If the progression of states illustrated in figure 2-3 constitutes a reconciliation of the objective with the subjective where that reconciliation can be described as a process, then a conceptualization of the perception of an object as a process becomes possible. It's interesting to note that, even though we've discussed the perception of an object as something inherently subjective, it's possible to conceptualize perception of (an object) as a process composed of two subprocesses: one subprocess starts from the object as it appears as expectation and ends at objective perception, and the other subprocess starts from subjective perception and ends at the object as it's observed as memory. The advantage of the subprocess approach to a conceptualization is two subprocesses temporally overlap to provide an intuitive visualization of how a state of subjective certainty and a state of objective uncertainty come together as the object observed as it appears in the now. Let's try something slightly different.

Conceptualizing perception of (an object) as a process is temporally consistent with a classic *input-processing-output* (IPO) scenario where input is the object-observed-as-it-appears as the object as it appears as expectation, output is the object-observed-as-it-appears as the object as it's observed as memory, and processing is perception of the object-observed-as-it-appears such that the object is observed as it appears in the now. In this conceptualization, the object observed as it appears in the now—a spatial reconciliation of the objective with the subjective—becomes a process where the now is defined when the object is recognized as observed as it appears.

Attempting to understand a conceptualization of the perception of an object as an IPO scenario reminds us that our primary interest is the phenomenological situation described as consciousness of (perception of (an object)). A more encompassing IPO scenario, consistent with the phenomenological situation illustrated in figure 2-4, requires the object-observed-as-it-appears to represent input, subjective perception to represent processing, and objective consciousness to represent output.

There is a sense of symmetry in describing consciousness as *consciousness of (an experience)* where a reconciliation makes it the objective expression of subjective perception of (an object). The output of interest becomes experience of the object observed as it appears in the now where experience is a process. This is informative, but, for the sake of conversation and consistency with our IPO scenario, we need to make a simplifying assumption. *Let's assume consciousness is temporal experience of the object observed as it appears in the now.* (Subjective experience of an object observed as it appears in the now is possible because objective consciousness has its own subject-object dichotomy conforming to a familiar morphology.) If consciousness is assumed to be temporal experience of the object perceived, then we can speculate regarding what makes the conceptualization of the phenomenological situation as an IPO scenario of value: consciousness provides information in the form of experience to an organism capable of making use of the information to achieve some goal.

Given what we know about the human organism, we sense that the phenomenological framework is capable of accommodating more than we've described. Could there be an additional phenomenon, with its own dynamics, that emerges from the intentional network on which the phenomenological framework is built? Is this emergent phenomenon capable of accounting for more than what's provided by a description of the apprehension of an object as a transformation or an IPO scenario? Can emergent phenomena in our conceptual frameworks, acting separately or together, accommodate what we understand as self-awareness? You can count on it.

The conceptual frameworks enable an organism's access to the relationship between an object and a subject from the first person perspective of something we've yet to define. The first person perspective of interest to us as human organisms is what's commonly referred to as the subject. The human subject, particularly in psychology, is a lot more elusive than you think. It's not the ego. Is our sense of self as a subject simply a matter of convenience—a tacit assumption masking an amalgamation of interacting processes—or is there something one can point to, somewhere in a situation conforming to a conceptual geometry, and claim it's a representation of either the subject of our experience or the subject of awareness of our experience? What we need is what we have, the map of a system capable of accommodating an internal observer.

Before we move on, there is a concern. The functional description presented as consciousness of (perception of (an object)) is an attempt to communicate the sense of dynamics of the phenomenological apprehension of an object as a transformation from something temporal, to something arguably spatial, then to something temporal again. We just have to keep in mind that because of the way we've defined consciousness, we've made it less of a function and more of an output. Functionally, consciousness is consciousness of something, but when we use consciousness to describe what it reveals, consciousness is temporal experience. The concern is defining consciousness as temporal experience, where objective consciousness obtains from the collapse of subjective perception of (an object), assumes a causal progression that appears inconsistent with our suggestion of an intentional network connecting objective consciousness of an experience with subjective perception of an object. Because the phrase is convenient, we'll continue to use consciousness of (perception of (an object)) to argue consciousness is the objective expression of subjective perception of an object. It's not the complete story but it will suffice.

In summary, perception transforms the temporal object-observed-as-it-appears to a spatial object observed as it appears in the now; this allows consciousness, the result of a further transformation of perception of (an object), to be conceptualized as our common sense notion of temporal experience of the object observed as it appears the organism can access as subjective experience of the perceived object. Consciousness and perception are also functionalities connected in the intentional network supporting the phenomenological framework in a manner requiring an explanation beyond the scope of this presentation.

(To help put the discussion in this section into perspective with respect to process philosophy, temporality, and individuation, you might find (Sherover 2003, chap. 6) of interest. For more on process philosophy, see (Mesle 2008; Rescher 1996).)

Chapter 3

Consciousness of (Perception of (an Object))

Introduction

In this chapter we'll present an interpretation of the map illustrated by the pyramids of Giza as the conceptualization of a phenomenological situation foreshadowed as consciousness of (perception of (an object)).[1] The interpretation will be made using the two lower levels of abstraction outlined in chapter two.[2] Keeping the interpretation at the lower levels of abstraction is essential to sidestepping a detailed analysis of the situation.

A convenient topic of departure is a geometric feature of the layout of the pyramids known as the *vanishing point*. Discussion of the vanishing point will be followed by discussion of a related geometric feature best described as the *cone of actualization* of consciousness. If the pyramids and their vanishing point represent subjectivity as perception of (an object), and if the cone of actualization and measures of consciousness the cone can accommodate represent objectivity, then we'll clearly need to combine the two representations to achieve a better understanding of how the map provided by the pyramids can be interpreted as consciousness of (perception of (an object)).

To make the presentation more intuitive, considerable effort will be invested in promoting a geometric interpretation of the phenomenological situation. We're going to argue that keeping the interpretation consistent with frameworks of the primary document necessitates reconciling measures representing objective consciousness with measures representing a *conceptualization* of subjective perception of (an object). The benefit of switching to a conceptualization of perception of (an object), the opposite of the approach taken in figure 2-4, will be its realization as measures situated along the axis of the cone of actualization–the same axis accommodating measures of consciousness. Conceptualizing subjective perception of (an object) as measures located at points along the axis of the cone of actualization of objective consciousness also requires the now to be conceptualized as one of those points.

All of the situated points will be integral to a proposed *super-pattern* that progresses from a state of subjective uncertainty to a state of objective certainty. In other words, we'll recognize recurring patterns in the conceptual geometry that can be combined to form a super-pattern, with a familiar morphology, encompassing all of the measures of consciousness of (perception of (an object)). The need to incorporate a super-pattern into the map will shed light on a prominent feature of the Great Pyramid.

Given the premise of this presentation, one should expect that the internal design of the Great Pyramid will contribute to our understanding of the relationship between perception of (an object) and consciousness. To that end, we'll offer an interpretation where the Great Pyramid presents details of the message encoded in the entire complex. Some of the details will create expectations regarding the phenomenological situation that can only be resolved in the primary document.

The Vanishing Point

Around 1979, Stephen Goodfellow had an idea regarding the pyramids of Giza. The pyramids appear to be positioned relative to each other in a manner that two circles passing through the southeast and northwest corners of the pyramids (see figure 1-1) might intersect at what can be described as a vanishing point.[3] Fortunately, John Legon produced a number of calculations and diagrams confirming the existence of the vanishing point. Two of the original figures have been reprinted by permission of the author in appendix 2. The vanishing point observation provides compelling evidence that the pyramids were built to conform to one design. You'd think a unifying design would be obvious by looking at them, but speculations regarding the design of the pyramids are typically molded to fit our current understanding of ancient history. (If you feel the need for more details regarding the layout of the pyramids, you might find Legon's analysis (Legon 2000B) supporting the plan of the Giza pyramids of interest.[4])

Figures 3-1 through 3-3 present our own illustrations of the vanishing point insight with additional features that may prove enlightening.[5] The only difference between the three figures is the extent to which they focus in to provide more detail. The vanishing point of interest is at point N. Since the pyramids represent perception of (an object)—a representation illustrating the progressive reduction in uncertainty of the subjective apprehension of an object as that object progresses from appearing as expectation to being observed as memory—the implication is the vanishing point at point N represents a point of zero uncertainty. If that's the case, then point N represents what can be called a certainty frame of reference.

Note that figure 3-1 introduces something new with respect to the original figure: there's an additional circle, centered at point A, passing through point K, where additional lines are drawn from point K to the centers of the base of the smallest and middle pyramids.[6] (The additional circle isn't shown in figures 3-2 and 3-3.)

The circle centered at point O in figure 3-1 is strategically located on a near forty five degree line drawn from the frame of reference at point A, to point M, so that points N and L of the smaller circle, centered at point O, are aligned with the larger circle, centered at point A. The larger concentric circles centered at point A suggest

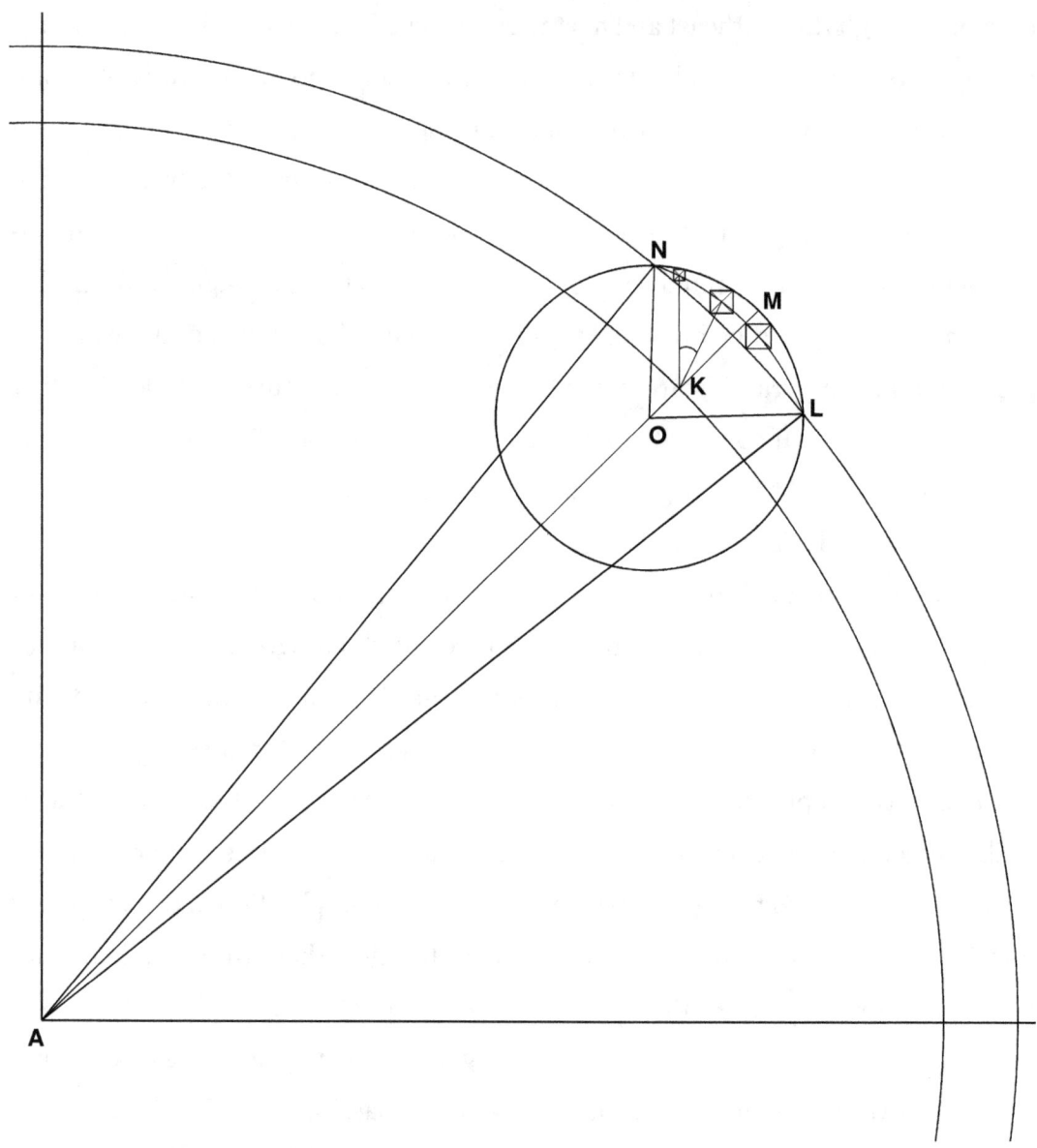

Figure 3-1

According to the Map

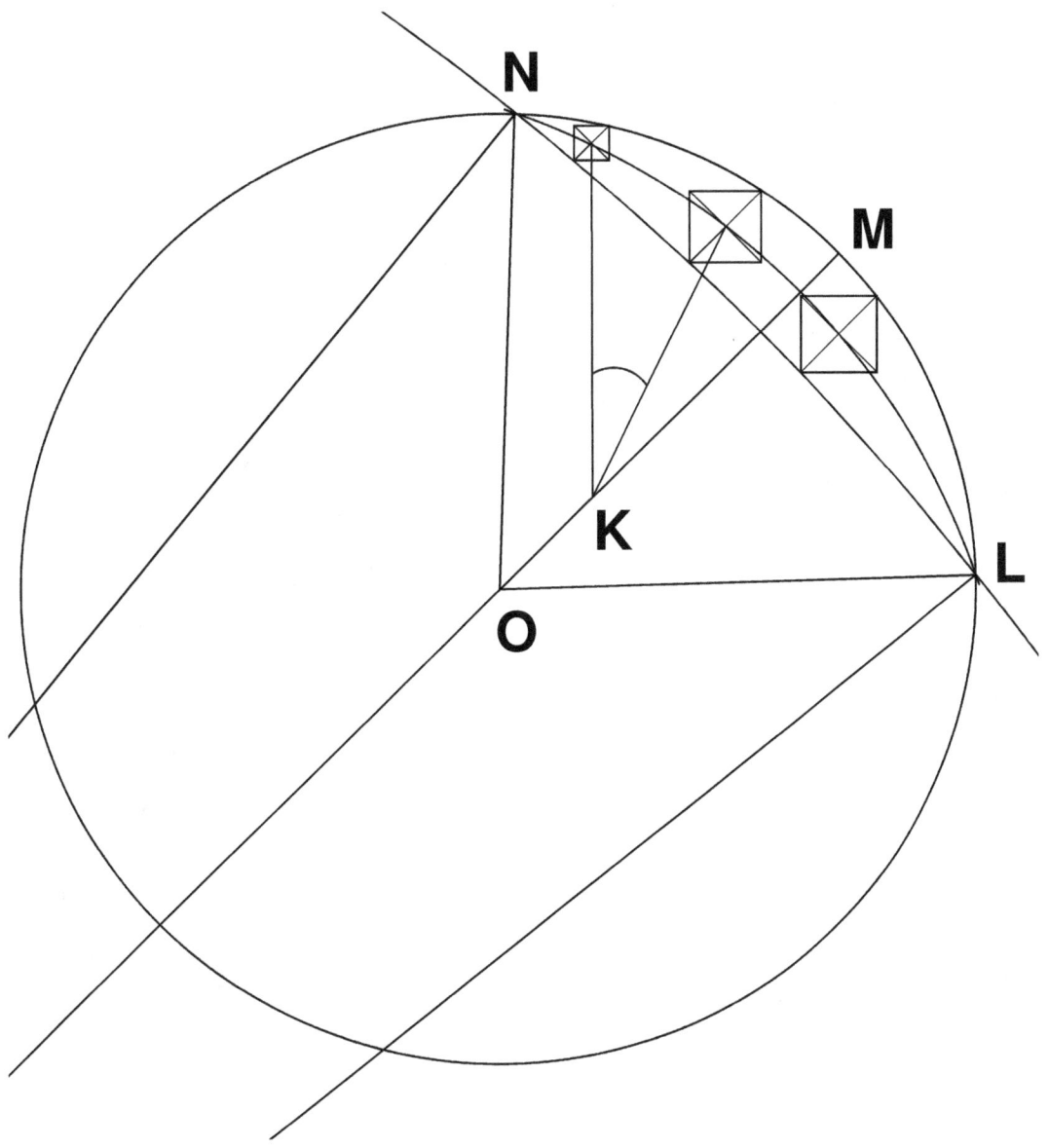

Figure 3-2

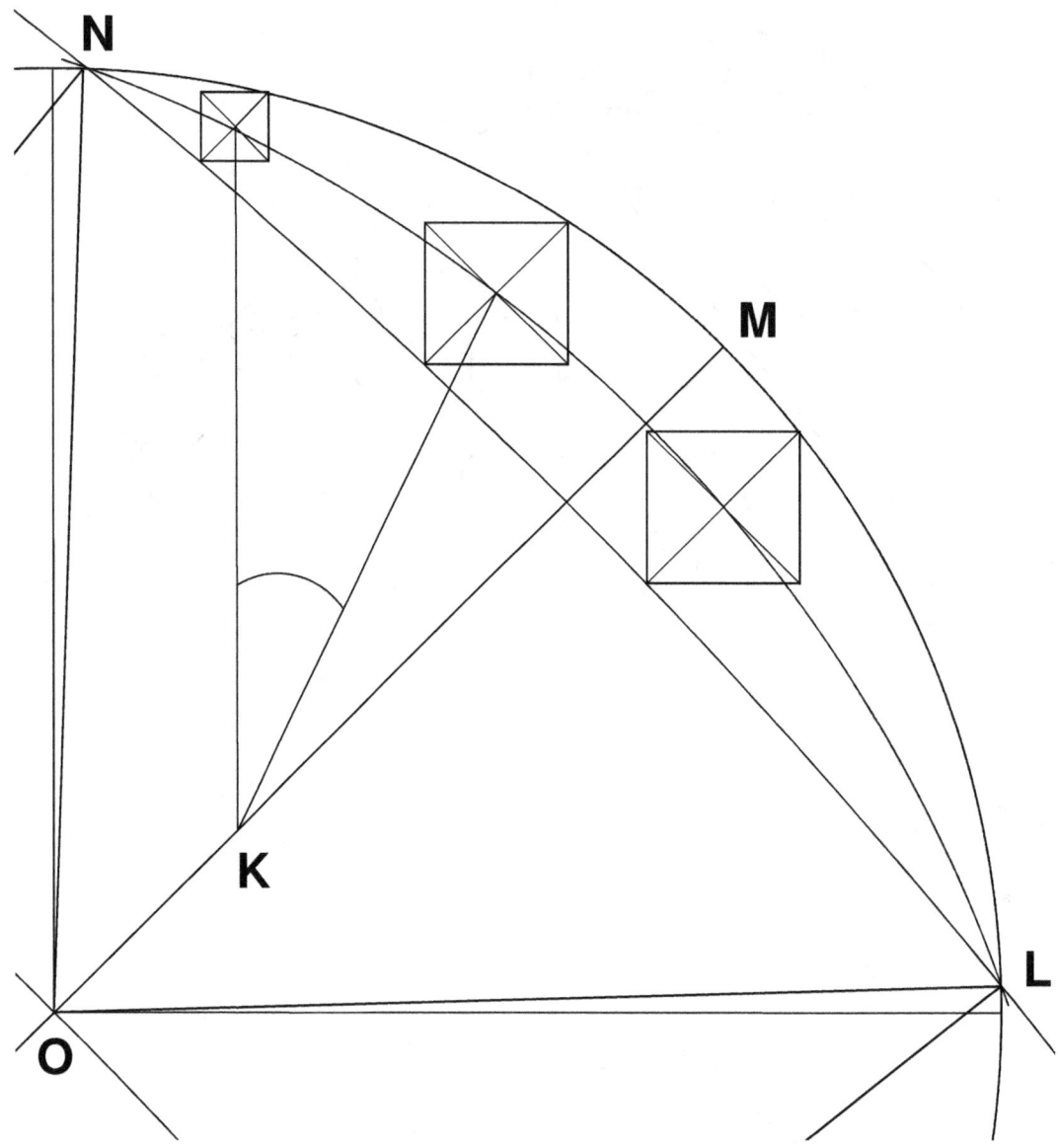

Figure 3-3

an underlying context that makes point K significant, but we needn't digress into those details to proceed. Figure 3-3 presents a close-up of figure 3-1 focusing on the relative positions of the pyramids with respect to two points of interest, one at point O and the other at point K. The more curious point of interest occurs at point K. As already mentioned, a vertical line drawn from point K intersects the exact center of the smallest pyramid, and there's also a line drawn from point K to the center of the middle pyramid. This line makes an angle of 25.99 degrees with respect to the vertical line. The value of the angle equals the value of the angle of inclination of the Ascending Passage of the Great Pyramid.[7]

An intuitive feel for what's communicated in the figures is important because the designers of the monument intended the vanishing point relationship as a means of suggesting a wider perspective. If all measures presented at the second level of abstraction are located in a multidimensional space representing uncertainty, then what's needed is a frame of reference from which to measure uncertainty. The frame of reference in this case is the certainty of point A in figure 3-1 representing a point of zero uncertainty. Do you see the problem? We've already claimed the vanishing point of the pyramids at point N is a certainty frame of reference with respect to perception of (an object), yet now we're claiming point A represents a certainty frame of reference at zero uncertainty. The situation needs to be put into perspective to resolve what could be interpreted as an inconsistency.

The certainty frame of reference at point N is part of the representation of subjectivity, and the certainty frame of reference at point A is part of the representation of objectivity. If figure 3-1 contributes to a map of the situation described as consciousness of (perception of (an object))—a relationship between subjectivity as perception of (an object), represented by three pyramids and their vanishing point, and objectivity as consciousness, represented by measures along the axis of a cone of actualization—then these certainty frames of reference can only be reconciled if we understand how the subjective becomes the objective. This is exactly what happens when the potential described as perception of (an object) is actualized as consciousness. *The easiest way to present the actualization of perception of (an object) as consciousness geometrically is if subjectivity as perception of (an object), represented by the progression of pyramids in the circle centered at point O, can be transformed into objectivity as consciousness, represented by one or more measures situated along the axis of the cone of actualization of consciousness drawn from point A to point M.* If the potential described as perception of (an object) can be actualized as consciousness in a manner suggested by figure 3-1, then how does this transformation help us when it comes to deciphering the map? Figure 3-1 is only the beginning of the story, a clue.[8] Investigation of the phenomenological situation supporting an argument that the pyramids present a map of cosmic significance can continue if a more comprehensive perspective is introduced.

Subjective Perception of (an Object)

Figure 3-4 illustrates what was briefly mentioned above as the cone of actualization of consciousness where four squares are drawn within that cone, spaced equidistant from each other, at points A, B, C, and D.[9] Since point A happens to be the apex of the cone of actualization, the square at point A can only be represented geometrically as a point—something you'll need to keep in mind. Each square, at a different level of uncertainty, appears to represent similar functionality suggested by the three pyramids they encapsulate representing perception of (an object). The bases of the three pyramids also present themselves as squares; however, they're not the squares of interest in this discussion. The squares of interest are those encompassing three pyramids at specific points along the axis of the cone; they represent the subjective in what will turn out to be an illustration of the objective. (As a matter of convenience, the axis of the cone of actualization in figure 3-4 and subsequent figures is drawn at a forty five degree angle with respect to the vertical and horizontal axes.)

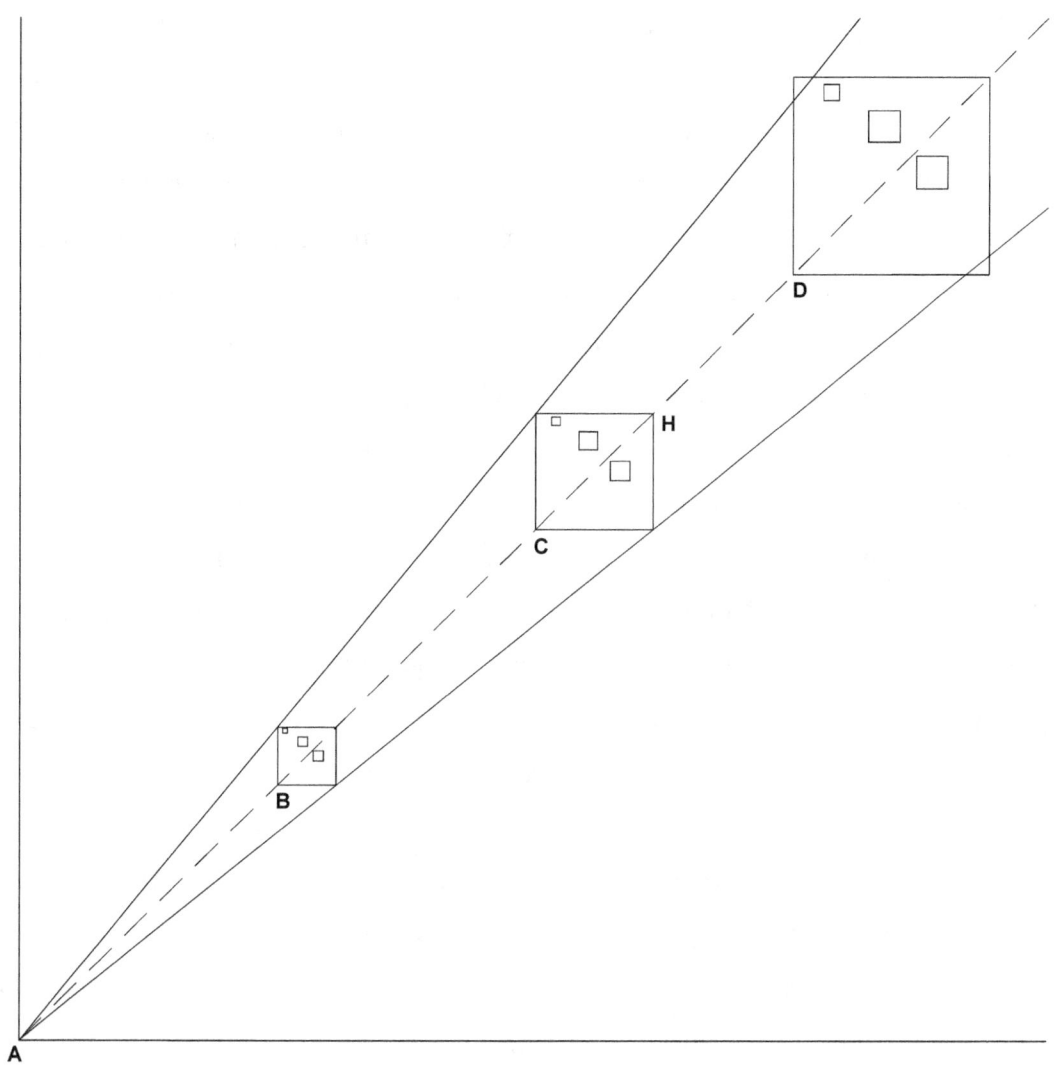

Figure 3-4

Consciousness of (Perception of (an Object))

If perception of (an object) represented by three pyramids at each of points A, B, C, or D in figure 3-4 can also be represented within a solid square at each point, then each square and the three pyramids it contains can be interpreted as a representation of subjective perception of (an object). Each representation of subjective perception of (an object) at a specific point along the axis of the cone of actualization of consciousness communicates a context for what the point represents. Though the size of each square must conform to the geometry of the cone of actualization upon which it's situated, it's reasonable to expect they communicate similar information.

Each representation of subjective perception of (an object), a solid square encompassing three pyramids, can be related to the difference in underlying measures of uncertainty between two points, situated along the axis of the cone, defining the square at that location. The relationship is illustrated in figure 3-4 by subjective perception of (an object) represented by three pyramids encompassed by the solid square bounded by points C and H. The locations of points C and H along the axis of the cone of actualization determine the area of the square and, consequently, some manifestation of subjective perception of (an object) the square and its enclosed pyramids represent.[10] The difference in underlying measures of uncertainty between points defining the square would be the same no matter where that particular square (size unchanged) is located along the axis of the cone of actualization, but, in this case, the size of the square accommodating perception of (an object) is uniquely determined by the geometry of the cone of actualization at point C. (It's tempting to try to compare subjective perception as three pyramids encompassed by a solid square at points A, B, C, and D in figure 3-4 to perception as subjectivity in figure 2-3–not yet; it's a comparison that will require figure 3-6 and additional information provided as the chapter progresses.)

Figure 3-5 shows the result of introducing a dashed square into figure 3-4 at points A, B, C, and D along the axis of the cone of actualization. The dashed square at point C isn't visible because it coincides with the solid square at that point.

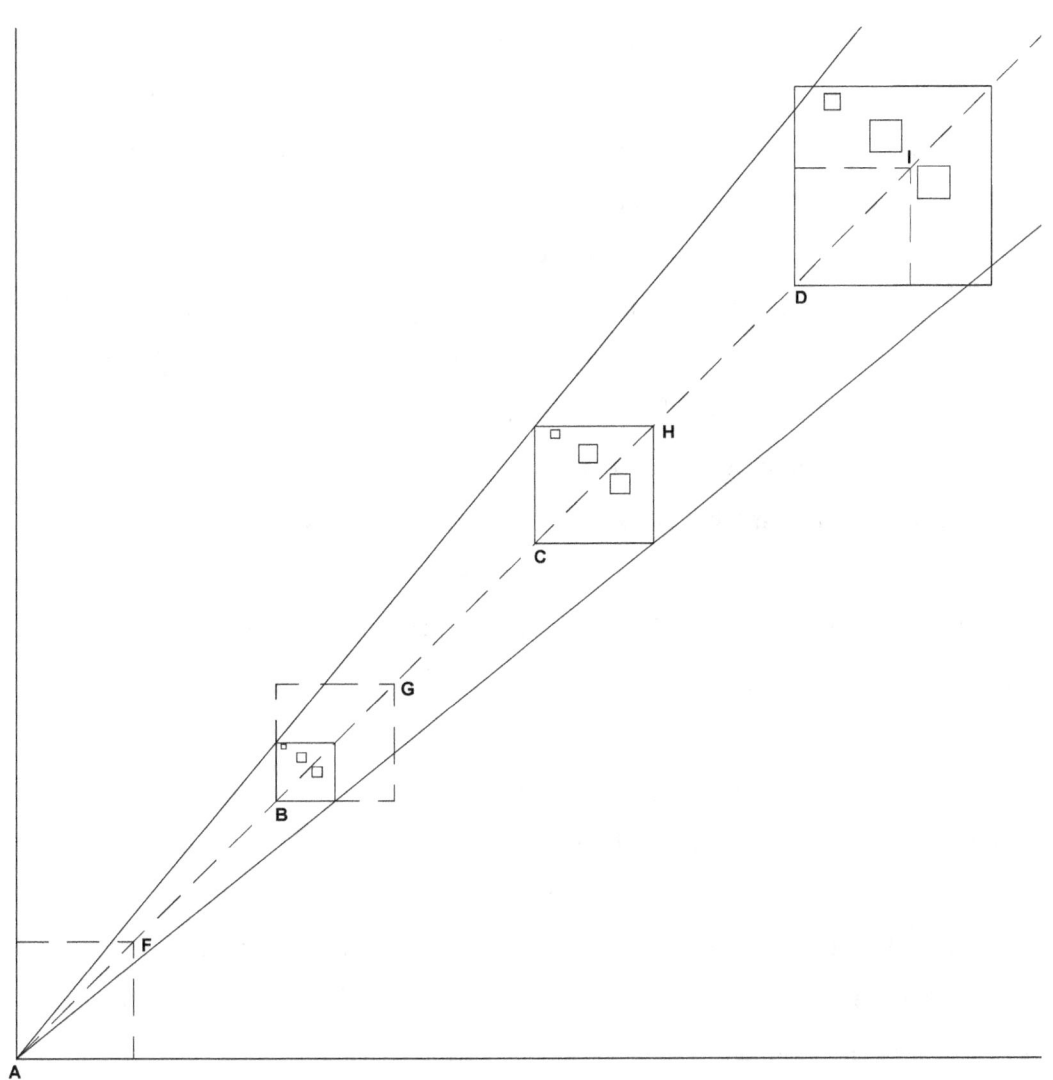

Figure 3-5

Consciousness of (Perception of (an Object))

The coincidence of dashed and solid squares at point C is a consequence of our choice of the dashed square's size, but it's not an arbitrary choice. Each dashed square is the same size because they all represent the same limit: a limit to the accommodation of perception of (an object) represented by the solid square at point C. The idea of a limit to the accommodation of perception of (an object) is consistent with constraints in the conceptual geometry that are independent of an interpretation of the geometry as a particular situation.

The dashed squares in figure 3-5 are for the purpose of presenting a visual comparison of subjective perception of (an object) at each of four points along the axis of the cone of actualization to subjective perception of (an object) at point C. As the presentation progresses, we'll be particularly interested in discovering if a limit to the accommodation of perception of (an object), we're claiming happens at point C, tells us anything about what the measure of consciousness at that limit represents.

So what is figure 3-5 telling us? The solid squares accommodating perception of (an object) at points A, B, and C are all smaller or equal in area when compared to the corresponding dashed square at each point. The implication is subjective perception of (an object) at those points along the cone of actualization conform to the limit represented by their dashed squares. By contrast, the solid square accommodating perception of (an object) at point D is greater in area than its corresponding dashed square, suggesting subjective perception of (an object) at that point along the cone of actualization doesn't conform to the limit represented by its dashed square. The distinction between what conforms and what doesn't conform in figure 3-5 will come to our attention again, in figure 3-13, when we discuss the Great Pyramid.

Let's look at the situation from a different perspective. Comparing the area of each solid square accommodating perception of (an object) to the area of the solid square at point C only makes sense if the solid square at point C represents a limit to the accommodation of perception of (an object). This description leads to an additional question: what's imposing that limit?

Consciousness of (Perception of (an Object))

The discussion in the previous section leads to speculation that a circle encompassing each of the three pyramids, similar to the circle centered at point O in figure 3-1, can be drawn at each of points A, B, C, and D along the cone of actualization in figure 3-5. The relationship is illustrated in figure 3-6. Figure 3-6 will be discussed in some detail but what's needed at the moment is an overview.

Let's start with an important observation: the vanishing point relationship within the circle centered at point O in figure 3-1 is the vanishing point relationship depicted at point B in figure 3-6. (There is a slight difference between the two representations due to the axis of the cone of actualization in figure 3-6 being drawn as a forty five degree line, among other things.[11]) Figures 3-7 and 3-8 focus on the vanishing point relationship at point B in figure 3-6 to present a clearer illustration of how what turns out to be one representation of consciousness of (perception of (an object)), as shown in figure 3-1, aligns with the progression in figure 3-6.[12]

We can propose that each circle and the three subjectively encapsulated pyramids it encompasses present consciousness of (perception of (an object)) as a subject-object dichotomy, where each circle represents objective consciousness. The center of each circle encompassing three pyramids is located along the axis of the cone of actualization at one of points A, B, C, and D. The implication is a measure at each point represents part of the expression of subjective perception of (an object) as objective consciousness.

If objective consciousness is represented by points A, B, C, and D in figure 3-6, then it's reasonable to suspect that consciousness will be compelled to collapse along the axis of the cone of actualization, from uncertainty at point D towards certainty at point A, as part of a reduction in uncertainty of the phenomenological

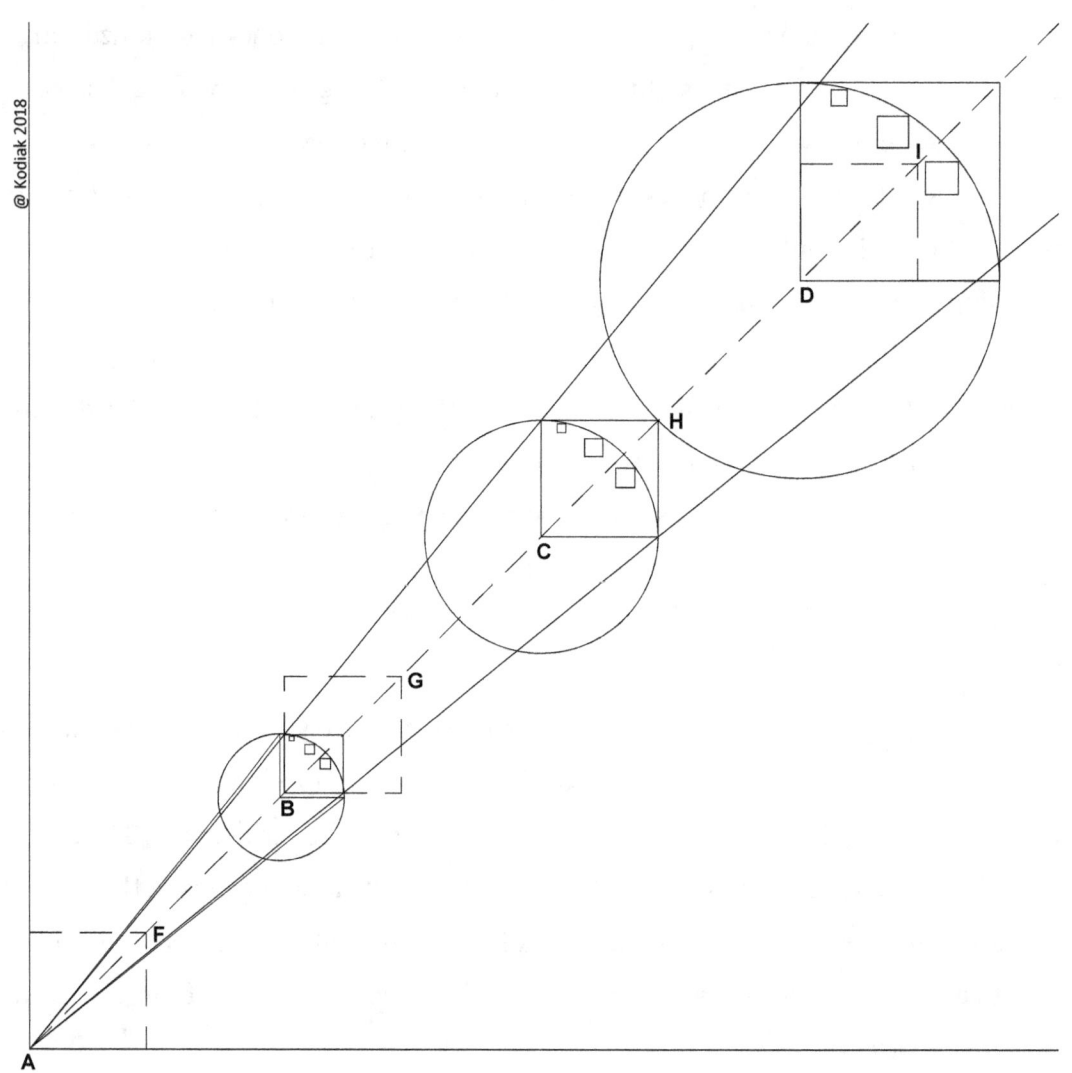

Figure 3-6

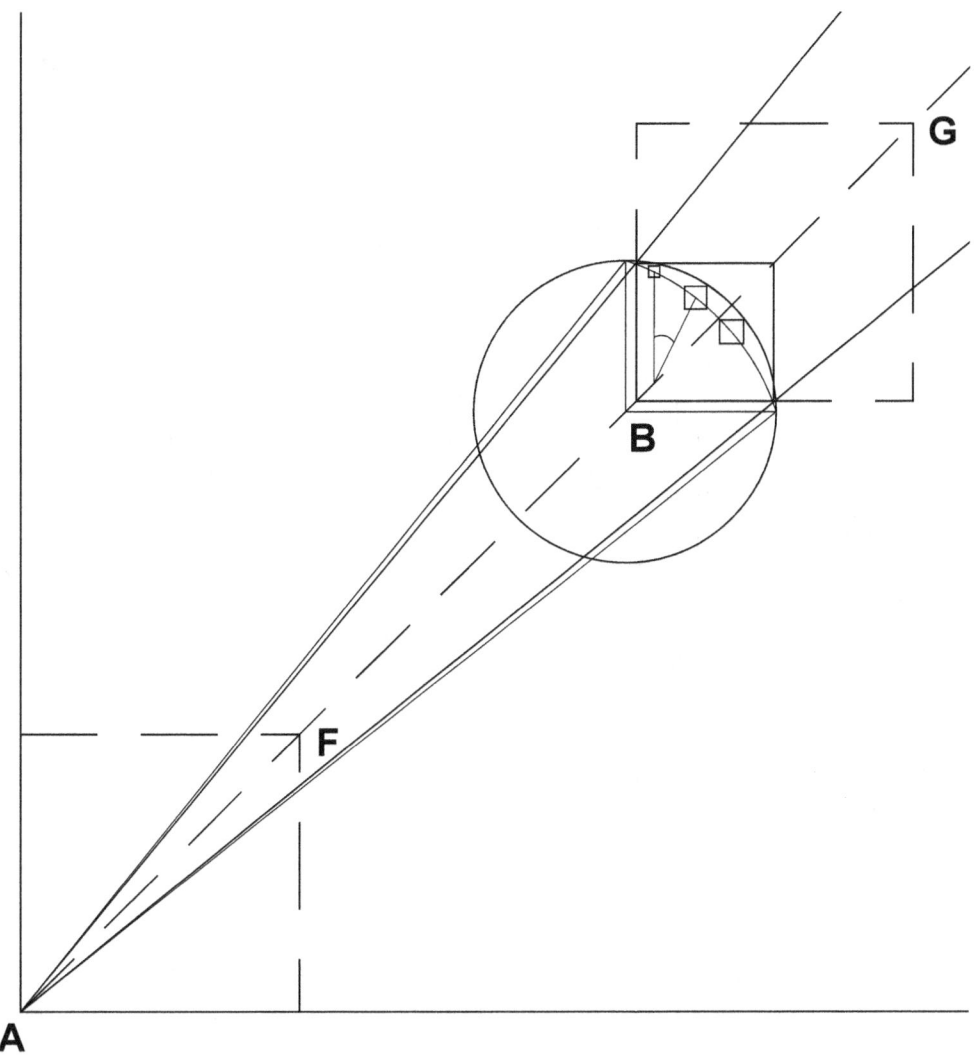

Figure 3-7

Consciousness of (Perception of (an Object))

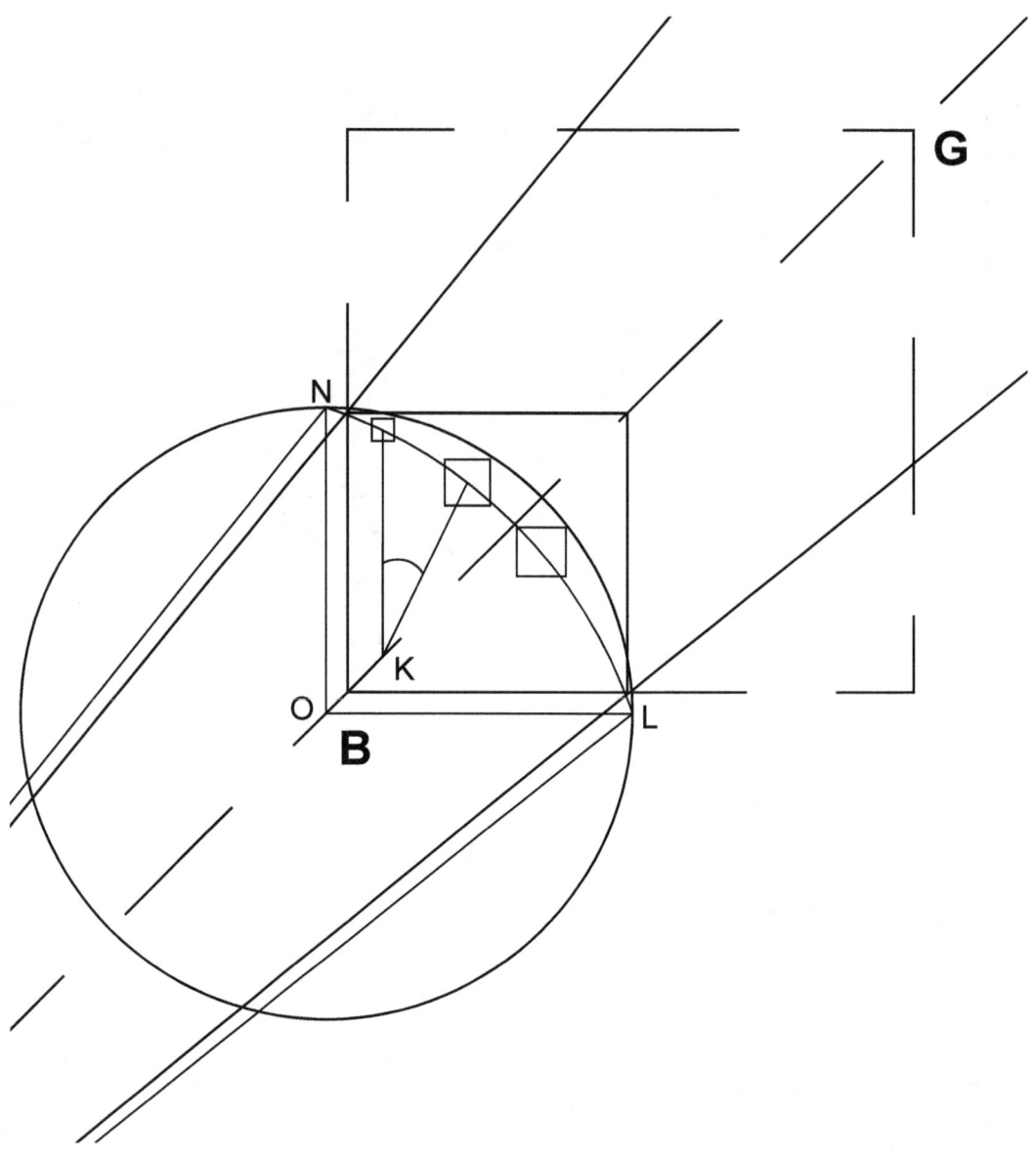

Figure 3-8

apprehension of an object. The assumed reduction in uncertainty of perception of (an object) and the assumed reduction in uncertainty of objective consciousness suggest consciousness of (perception of (an object))–represented geometrically by three subjectively encapsulated pyramids encompassed by circles centered at four points along the axis of the cone of actualization–*collapses* to measures of objective consciousness at points A, B, C, and D.[13] If the collapse is simultaneous at all four points, then all that remains in figure 3-6 is consciousness represented by points A, B, C, and D. Four actualized measures of consciousness are situated along the axis of the cone of actualization at points corresponding with locations of the center of each circle, locations *prescribed* by an underlying measure of uncertainty at the first level of abstraction. The collapse reflects a change in representation, from subjective perception of (an object) to objective consciousness, triggered by an event enabled by the intentional network the phenomenological situation is built upon–an event precipitated by the object being observed as it appears in the now.

The idea that three pyramids representing the perception of an object can be presented as subjective potential that can be actualized as objective consciousness is a big part of what figure 3-6 is communicating. *The idea is illustrated at all four points representing consciousness because figure 3-6 is communicating, in the only way it can, the dynamics of the phenomenological situation that represents one actualization at all four points.* A transformation and its direction are suggested when the situation is described as consciousness of (perception of (an object)). To make the transformation of perception of (an object) to consciousness easier to comprehend, we've chosen to present consciousness as an expression of objectivity and perception of (an object) as an expression of subjectivity such that the two apprehensions relate to each other as a subject-object dichotomy. Later, a point representation of perception of (an object) will be introduced as a conceptualization. That conceptualization will allow us to discuss the phenomenological situation from a different perspective. Fortunately, it's a point representation figure 3-6 already illustrates.

Why does the phenomenological apprehension of an object require a transition from perception to consciousness, and what exactly is implied by describing the transition as a collapse? The implication is subjectivity as the perception of an object, when that object is observed as it appears in the now, is expressible objectively as consciousness. A change in representation is possible because the measures of perception and consciousness are intimately connected by a network of intentionality that has evolved to facilitate the phenomenological apprehension of an object. The proposed collapse of perception of (an object) implies conditions allow for measures of subjectivity to be abandoned in favor of measures of objectivity. A transformation occurs. There's only what remains after an object is perceived to be observed as it appears—only consciousness. Why bother? *Objectivity as consciousness in the situation we're describing is free of perception and what perception requires; consequently, a different subjective frame of reference becomes possible.*

Is it possible that objective consciousness is more personal than subjective perception? Can consciousness and its subjective frame of reference offer a more personal presentation of the perception of an object? Because of its morphology—something we'll discuss shortly—objective consciousness enables the experience of sensations associated with the object observed as it appears in the now.

(Even though it's not obvious in figure 3-6, think about why changing the subjective frame of reference from a measure of subjective certainty of perception to a measure of subjective certainty of consciousness might be desirable. That change will be implied in figure 3-13 when the design of the Great Pyramid is discussed. Figure 3-13 will present two pyramids: the Great Pyramid is the primary focus; however, the smaller pyramid encompassed by the Great Pyramid, illustrated using dashed lines, suggests an intimate connection between perception and consciousness we'll need to explore.)

Returning to figure 3-6, it seems plausible that perception of (an object) is capable of collapsing to consciousness at points A, B, C, and D. Just keep in mind that each of the four points represents a different but related contribution to consciousness. The measures representing consciousness at points A, B, C, and D can be related to each other by the same morphology, defined by the relationship between states of objective and subjective certainty and states of objective and subjective uncertainty, discussed with respect to perception of (an object) in chapter two. In chapter two, we suggested that an understanding of consciousness is possible without loss of our familiar sense of egocentricity. *The measure of consciousness at point B is what gives consciousness something resembling egocentricity even though consciousness represents objectivity with respect to the subjectivity of perception of (an object).* A sense of egocentricity is consistent with the observation that if the measure of consciousness at point A in figure 3-6 represents a state of objective certainty and the measure of consciousness at point D represents a state of subjective uncertainty, then the order of progression from uncertainty to certainty, discussed in chapter two, requires the measure of consciousness at point B to be a state of subjective certainty; that state makes point B the subjective frame of reference of consciousness represented by points A, B, C, and D. None of the four points represent the *subjective* now. Why? The four points representing objective consciousness actualize the potential represented by subjective perception of (an object) when that object is observed as it appears in the now. The subjective now's contribution is limited to bringing the object as it's observed and the object as it appears together as the object observed as it appears.

In summary, if what's illustrated in figure 3-6 is a representation of the actualization of subjective perception of (an object) as measures of objective consciousness, situated at well-defined points along a progression from uncertainty to certainty, then it's likely that this figure might be communicating information about a situation described as consciousness of (perception of (an object)).[14] (It's tempting to try to improve the flow of the presentation by dropping the word subjective from the phrase 'subjective perception of (an object)' in favor of assuming perception of (an object) is always subjective with respect to consciousness of (perception of (an object)), but it's a simplification that may not be worth the lost in clarity. Although we've simplified the terminology where it seems appropriate, don't be surprised if you see the phrase again.)

This is a convenient point in the presentation to note that we're willing to discuss the particulars of how perception relates to the object-observed-as-it-appears, a relationship between measures described as the object as it's observed as perception and the object as it appears as perception with respect to the object observed as it appears in the now, yet we won't have much to say regarding measures describing consciousness at points A, B, C, and D. Discussing consciousness would be a bit more involved than what's represented by points A, B, C, and D, and it's a discussion that isn't essential to achieving our primary goals.

The Collapse to Certainty

One of the ideas presenting itself in figure 3-6, particularly when the situation is contemplated in terms of a progression from uncertainty to certainty, is the possibility that consciousness of (perception of (an object)) will not remain as depicted along the cone of actualization; it collapses to certainty at the certainty frame of reference represented by point A. Based on our current argument, one would expect perception to collapse to consciousness prior to the collapse of consciousness to certainty, yet it's entirely possible that if they collapse, the dynamics of that collapse could be anywhere between sequential and simultaneous. The easiest way to make the possibility of collapse of consciousness of (perception of (an object)) to certainty more intuitive is to manipulate the geometric representation of the situation presented in figure 3-6 to create a visual approximation of a path the collapse might take. As shown in figure 3-9, consciousness of (perception of (an object)) at point D is shifted in the direction of certainty to point C so that points C and D align and the circles centered at points C and D become concentric. The situation

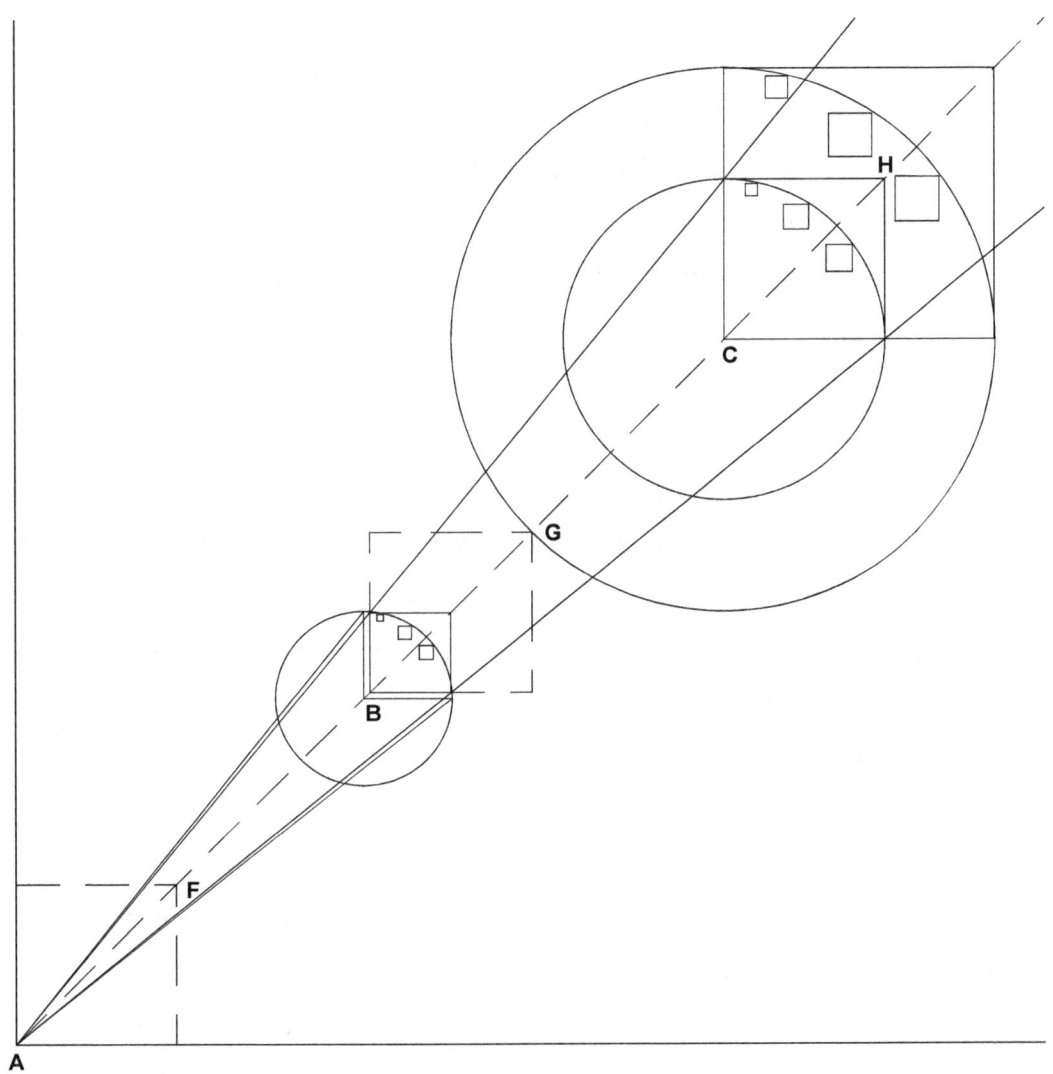

Figure 3-9

Consciousness of (Perception of (an Object))

in figure 3-9 gives the impression that what's collapsing to certainty is not only the reduction in size of each of the three pyramids representing perception—pyramids encompassed by each circle representing consciousness—it's also the coordinated collapse of consciousness in the direction of certainty suggested by the shift of the circle centered at point D to point C.

Figure 3-10 shows what might conceivably be the next step in the collapse of consciousness of (perception of (an object)) to certainty. In this figure, both concentric circles centered at point C in figure 3-9 and perception of (an object) each circle encompasses are now centered at point B. There are three concentric circles providing a compelling illustration of the collapse of consciousness of (perception of (an object)).[15]

The last step in the collapse of consciousness of (perception of (an object)) to a certainty frame of reference at point A isn't shown in a separate figure simply because there wouldn't be much to show. All three concentric circles at point B in figure 3-10 and perception of (an object) each circle encompasses would be centered at point A: three pyramids encompassed by a square and a circle would be actualized as consciousness at points A, B, C, and D *coincident* with the certainty frame of reference at point A, which suggests all of consciousness of (perception of (an object)) illustrated in figure 3-6 collapses to a single point of certainty at point A. Yeah, there's a concern.

The essence of the argument is potential revealing itself as perception of (an object), contingent on the perceived object being observed as it appears, becomes actualized as consciousness; however, when it comes to what happens next, it's not clear that consciousness collapses to certainty at point A—even if it's compelled to do so—or that such a collapse describes what we experience. The outcome of interest is probably represented by figure 3-6. In that figure, perception of (an object) is actualized as consciousness at points A, B, C, and D such that four measures of consciousness are related to each other in a manner that's revealing. The other possibility is consciousness of (perception of (an object)) does collapse to certainty at

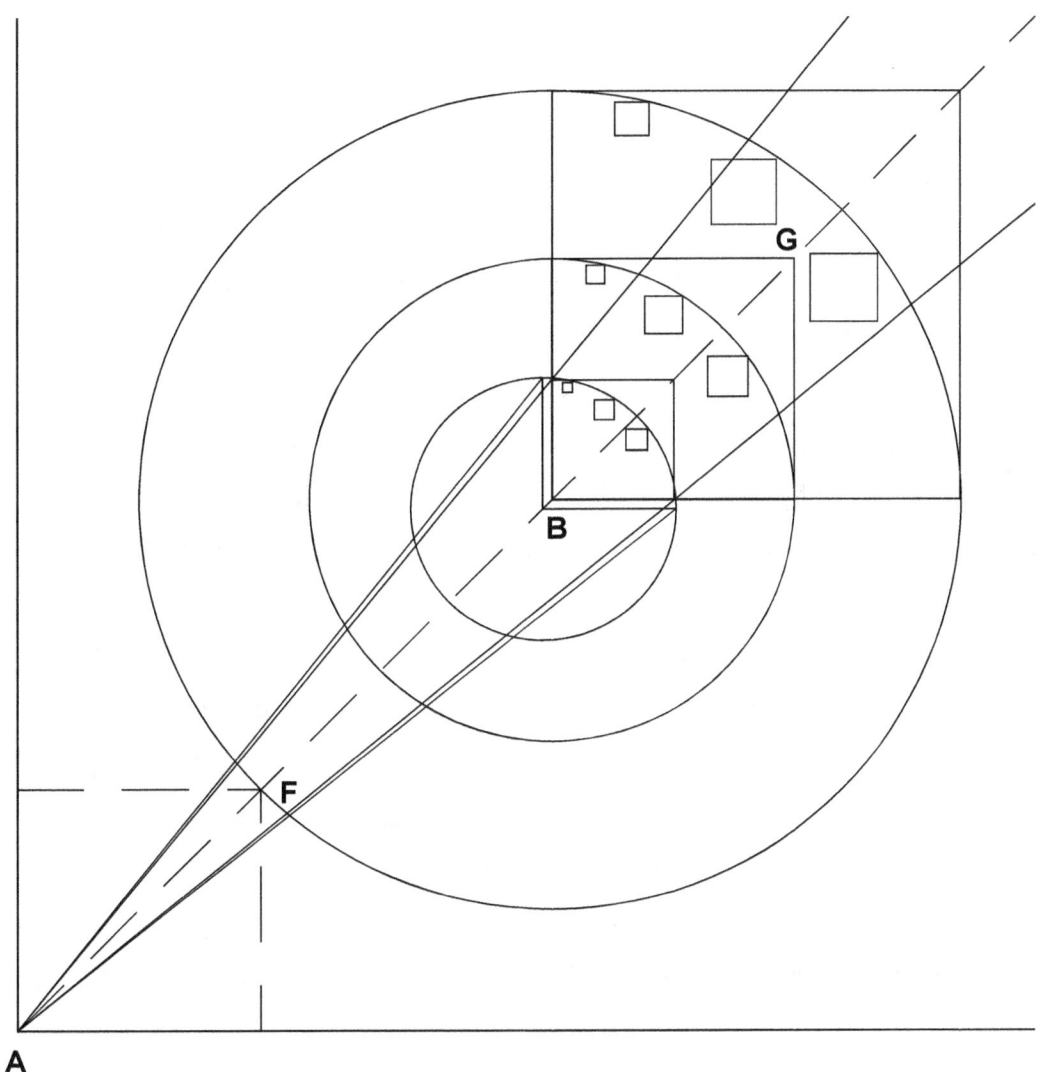

Figure 3-10

Consciousness of (Perception of (an Object))

point A on an occasion that can range from rare to continuous, but we as organisms may not be directly aware of the collapse.[16]

Conceptualizing Perception of (an Object)

Subjective perception of (an object) is presented as three pyramids encompassed by a solid square at each of points A, B, C, and D along the axis of the cone of actualization. There are four representations, one at each measure of consciousness, for a reason: representations of subjective perception of (an object) encompassed by a circle representing objective consciousness are instrumental in visualizing the claim that consciousness can be understood as consciousness of (perception of (an object)). Unfortunately, the approach isn't as insightful when it comes to comparing the morphology of perception of (an object) to the morphology of its actualization as consciousness. *What's needed is a conceptualization of perception of (an object) along the axis of the cone of actualization allowing a more direct comparison of subjective perception to objective consciousness at points A, B, C, and D.*

Let's suppose points F, G, and H in figure 3-6 conceptualize the inner horizon of perception of (an object) where, as illustrated in figure 2-3, the object is perceived to be observed as it appears. In this conceptualization, one can claim the object is observed as it appears in a now at point G where point G is situated between points F and H representing perception. On the basis of location, a now at point G is a strong candidate for a now defined by an object observed as it appears whose perception collapses to consciousness represented by points A, B, C, and D.

So what's to be gained by introducing a conceptualization of perception of (an object) as points along the axis of the cone of actualization? As it turns out, the framework in the primary document representing the situation described as consciousness of (perception of (an object)) utilizes a now analogous to a point along the axis of the cone of actualization in figure 3-6—a now defined by an object observed as it appears conceptualized at point G. The framework requires point G to be situated between perception conceptualized at points F and H where point F represents the object observed as perception. Because the realization of a framework in the primary document is the ultimate goal, we're inclined to introduce interpretations consistent with that framework. Introducing the conceptualization of perception of (an object) as points F, G, and H along the axis of the cone of actualization will also turn out to be convenient to fully appreciating insights presented in the remaining sections of this chapter. (Now you can compare perception in figure 3-6 to perception illustrated in figures 2-3 and 2-4.)

Why didn't we simply start by conceptualizing perception of (an object) as points along the axis of the cone of actualization? Suppose we hadn't started with the idea that a solid square encompassing three pyramids represents subjective perception of (an object). Suppose we'd started with the cone of actualization of consciousness and four dashed squares of equal size located at points A, B, C, and D (similar to what's shown in figure 3-5) where the dashed squares align with points F, G, H, and I positioned along the axis of the cone. Without insight provided by the solid squares and the pyramids they encompass, one could conceivably advance an argument that points A, B, C, and D are actualized measures of consciousness and points F, G, H, and I are conceptualized measures of perception of (an object), but the *idea* that actualized measures of consciousness result from the collapse of potential represented by perception of (an object) would be impossible to communicate—think about it. The progression towards certainty of the cone of actualization of consciousness and the dashed squares representing some limit to accommodation of subjective perception of (an object) simply aren't sufficient to communicate what makes the phenomenological situation described as consciousness of (perception of (an object)) compelling. What makes the situation compelling is the idea that perception of (an object), under the right conditions, collapses to consciousness as consciousness itself is compelled to collapse to certainty. We're talking about a dynamic that can be attributed to an underlying network of intentionality required for apprehension of an object as it progresses from appearing to being observed. A key contribution of the pyramids of Giza is this particular insight—insight figure 3-6 communicates.

(Is it possible that points G, H, and I in figure 3-6 conceptualize the inner horizon of perception of (an object)? This conceptualization implies the object is observed as it appears in a now at point H where point H is situated between points G and I representing perception. The scenario making the most sense for this additional possibility is when a now at point H and the now at point G represent the same now at different points in a progression from uncertainty to certainty of consciousness of (perception of (an object)) illustrated in figure 3-6. A now at point G is assumed to be associated with the collapse of perception of (an object) to consciousness represented by points A, B, C, and D in a manner that the measure of objective certainty is the certainty frame of reference at point A and the measure of subjective certainty is at point B. We can speculate that consciousness and perception can also be represented one step earlier and further out in the direction of uncertainty: point B becomes a measure of objective certainty and an objective frame of reference, the measure of subjective certainty is realized at point C with respect to point B, and so on. The problem with this speculation is the conceptual geometry upon which the points are mapped dictates a relationship between the object observed as it appears defining the now at point G and the certainty frame of reference defined by point A the proposed shift violates. If the conceptual geometry constrains the relationship, then we needn't speculate any further.[17])

A Moment of Reflection about a Moment

Even though our description of the now is limited to the object observed as it appears in the now, represented by the middle pyramid in figure 2-3 or conceptualized by point G in figure 3-6, if you step back and reflect on the situation described as consciousness of (perception of (an object)), you get the impression—even without knowing particulars of what consciousness represents–that the entire situation is dedicated to representing or communicating what's transpiring now. This impression is reinforced by an outer horizon, the object as it appears as expectation and the object as it's observed as memory, beyond the field of perception of the object observed as it appears in the now.

Once an object is observed as perception as it appears as perception and the perception of an object, observed as it appears in the now, collapses to objective consciousness, it seems probable that any sense of the now in objective consciousness is directly related to that transformation. What happens if consciousness resulting from the collapse of perception of (an object) contains no sense of temporality beyond the now? [18] What would be required to achieve consciousness of time?

Tell Tale Patterns

If we pursue a different approach to understanding figure 3-6, keeping in mind subjective perception of (an object)–three pyramids encompassed by a solid square–is represented at point A as a point, a repeating pattern becomes recognizable: points A, F, B, and G represent one occurrence of a pattern that repeats itself at points B, G, C, and H and repeats itself again at points C, H, D, and I. When their points are considered in reverse order, each of the patterns is consistent with a progression from subjective uncertainty to objective certainty. Points associated with objectivity, points A, B, C, or D, are actualized measures of consciousness and points associated with subjectivity, points F, G, H, or I, are conceptualized measures of perception of (an object). If the three patterns are tied together by a more comprehensive pattern described by points A, G, C, and I, then these patterns could contribute to a patterned progression from uncertainty to certainty expressed in the form of a relationship between objective measures of consciousness and subjective measures of perception of (an object).

(The goal in this section is the introduction of a set of patterns that can be of benefit in understanding the construction of the Great Pyramid. We're also limiting the discussion to patterns we can claim are intuitive representations of a subject-object dichotomy. The patterns mentioned above are presented primarily as an example that happens to be visually easy to grasp.)

There are patterns in the situation figure 3-6 illustrates with the potential of being more revealing. To identify them and appreciate how they're interrelated, it's necessary to revisit an observation made earlier with respect to subjective perception of (an object):

We've established that a solid square encompassing three pyramids represents subjective perception of (an object) at each of points A, B, C, and D along the axis of the cone of actualization. If a dashed square at each point represents a constraint on how much perception of (an object) can be accommodated, then what is the nature of that constraint? To answer this question, we need to rethink the comparison of perception of (an object) accommodated by each solid square to perception of (an object) accommodated by the dashed square at the same point. The difference in area between the solid and dashed squares at each point indicates the extent to which perception of (an object) accommodated by the area of the solid square could be accommodated by the area of the dashed square, where the area of each dashed square—since they're all the same size—is dictated by one position along the axis of the cone of actualization.

Two aspects of the situation illustrated in figure 3-6 come into focus: first, at each point A, B, C, and D along the cone of actualization, it's likely that perception of (an object) accommodated by the area of the solid square is compared to the same constraint because the constraint represented by the area of the dashed square at each point is invariant; second, increasing subjectivities of perception of (an object)—perception of (an object) accommodated by the solid squares at points A, B, C, and D—*correlate* with a linear measure of uncertainty at the first level of abstraction (made with respect to the certainty frame of reference at point A) that progresses to different points along the axis of the cone but progresses in a manner that each occurrence of the measure encompasses increasing amounts of uncertainty as it encompasses more of the cone of actualization. That measure of uncertainty is understood at the second level of abstraction as the uncertainty of observation: a measure represented by the distance from point A to point A in the trivial case associated with the solid square (represented by a point) at point A, the distance from point A to point B associated with the solid square at point B, the distance from point A to point C associated with the solid square at point C, and the distance from point A to point D associated with the solid square at point D. *In each case, a corresponding pattern of measures—let's call it a sub-pattern—has the same morphology of objective certainty and uncertainty in a relationship with subjective certainty and uncertainty. Each sub-pattern progressively reaches further out in the direction of uncertainty as a different manifestation of points along the axis of the cone of actualization each time the uncertainty of observation increases.*[19] Let's make use of figure 3-6 to examine the four sub-patterns in more detail.

In the trivial case, subjective perception of (an object) at point A is a point, and the uncertainty of observation from point A to point A is zero. Because perception of (an object) is accommodated by the solid square at point A as a point, its area is obviously less than the area of the dashed square at that same point. In other words, perception of (an object) accommodated by the solid square at point A is within the constraint represented by the dashed square. In this case, point A *is* the sub-pattern of interest. A more intuitive relationship reveals itself as we move out along the axis of the cone of actualization to point B.

With respect to perception of (an object) accommodated by the solid square at point B, the sub-pattern of interest is one we've already introduced at the beginning of this section as points A, F, B, and G. In this sub-pattern, uncertainty of the frame of reference is represented by the distance from point A to point F, the uncertainty of measurement is represented by the distance from point A to point

G, and the uncertainty of observation—our primary interest—is represented by the distance from point A to point B.[20] The area of the solid square at point B is related in a well-defined manner to the uncertainty of observation (the distance along the axis of the cone of actualization from point A to point B) associated with the sub-pattern represented by points A, F, B, and G; but the key insight requires noting that the area of the dashed square located at point B is related to the uncertainty of observation (the distance along the axis of the cone of actualization from point A to point C) associated with a different sub-pattern that belongs to consciousness represented by points A, B, C, and D. *The difference in area between the solid and dashed squares at point B represents what can be conceptualized as the difference in the amount of perception of (an object) each sub-pattern can accommodate at its uncertainty of observation.* (Everything said with respect to the dashed square located at point B is also true of the dashed squares at points A, C, and D because each dashed square represents the same constraint.)

The above insight suggests the solid and dashed squares at point C are coincident because both squares manifest at the same uncertainty of observation (the distance from point A to point C) associated with consciousness represented by points A, B, C, and D. At point C we're comparing the amount of perception of (an object) that can be accommodated by the solid square at the uncertainty of observation of consciousness with itself.

Finally, the solid square at point D accommodates perception of (an object) at the uncertainty of observation (the distance from point A to point D) of a sub-pattern that's obviously too big to present in figure 3-6. We can speculate that it's a sub-pattern representing consciousness of (perception of (an object)) containing measures of consciousness and conceptual measures of perception of (an object). Figure 3-14 will be used at the end of this chapter to identify the pattern as points A, G, D, and J. Perception of (an object) the sub-pattern can accommodate, dictated by the area of the solid square at point D, is greater than perception of (an object) accommodated by the area of the dashed square at point D; we're beyond the limit of the amount of perception of (an object) that can be accommodated by consciousness represented by points A, B, C, and D.

Consider the four sub-patterns discussed above with respect to the uncertainties of observation represented by the distance from the certainty frame of reference at point A to each of points A, B, C, and D: Sub-patterns associated with uncertainties of observation represented as distances ending at points A and C are both sub-patterns of consciousness containing only points representing measures of consciousness. In the former case, consciousness is the trivial sub-pattern represented by the certainty frame of reference at point A where distance along the cone of actualization reflecting the uncertainty of observation is obviously zero, and, in the latter case, consciousness is represented by the familiar points A, B, C, and D. Sub-patterns associated with uncertainties of observation represented as distances ending at points B and D, on the other hand, are both sub-patterns of consciousness of (perception of (an object)) where each sub-pattern relates points representing conceptual measures of perception of (an object) to points representing actualized measures of consciousness. (Keep in mind that the entire situation illustrated by figure 3-6 is described as consciousness of (perception of (an object)), a situation encompassing all sub-patterns. Referencing a sub-pattern or the entirety of the situation illustrated by figure 3-6 depends on the context. We'll try to make an appropriate qualification going forward to make the intent more obvious.)

If we associate the two sub-patterns representing consciousness with objectivity and the two sub-patterns representing consciousness of (perception of (an object)) with subjectivity, we've created exactly what's required for a super-pattern connecting the four sub-patterns that may lead to a better understanding of how consciousness of (perception of (an object)) is compelled to collapse to certainty at the certainty frame of reference at point A. The collapse is achieved via a progressive reduction in the uncertainty of observation. The super-pattern requires associating the larger sub-pattern representing consciousness of (perception of (an object)) with subjective uncertainty and the smaller sub-pattern representing consciousness of (perception of (an object)) with subjective certainty. The smaller sub-pattern is identified as points A, F, B, and G, and although the larger sub-pattern is too big to illustrate in figure 3-6, it's identified as points A, G, D, and J in figure 3-14. The super-pattern can be completed if we associate consciousness represented by its collapse to certainty at point A with objective certainty and associate the sub-pattern representing consciousness at points A, B, C, and D with objective uncertainty.[21]

The progression of this super-pattern from the sub-pattern representing subjective uncertainty to the sub-pattern representing objective certainty makes the relationship between each of the sub-patterns intuitive, with the added bonus that comparing the area of each solid square located at points A, B, C, and D to the constraint represented by the corresponding dashed square—at the same location—becomes intuitive as a comparison of subjective perception of (an object) accommodated at the uncertainty of observation of each sub-pattern to subjective perception of (an object) accommodated at the uncertainty of observation of the sub-pattern defining consciousness represented by points A, B, C, and D.

The immediate question that comes to mind is what's so important about comparing perception of (an object) accommodated by the solid square at the uncertainty of observation of each sub-pattern to perception of (an object) accommodated by the solid square at the uncertainty of observation of the sub-pattern defining consciousness? The answer suggesting itself is there's no point in perceiving more than can be cognized. A limit to the amount of perception that can be cognized is suggested by subjective perception of (an object) represented by the solid square encompassing three pyramids at the uncertainty of observation, at point C, associated with consciousness represented by points A, B, C, and D. The challenge will be to figure out what more if anything this constraint tells us beyond what appears obvious.[22]

Let's Take Our Bearings

We might benefit from taking stock of our understanding of the phenomenological situation by presenting a summary of what's illustrated in figure 3-6 from a slightly different perspective.

The multidimensionality of consciousness of (perception of (an object)) can be suggested by starting with a progression of circles representing consciousness where each circle is situated along a cone of actualization correlating uncertainty with dimensionality in such a way that the circles are centered at points A, B, C, and D. The location of each point is prescribed by the underlying conceptual geometry of a generic situation.

Each circle representing consciousness encompasses three pyramids that are also encompassed by a solid square. The three pyramids encompassed by a solid square represent subjective perception of (an object). The intent of the geometric relationship between representations at points A, B, C, and D is to suggest a *network of intentionality* that enables potential represented by subjective perception of (an object) to be actualized as objective consciousness. Consciousness is subsequently represented by four measures situated along the axis of the cone of actualization at points A, B, C, and D. To avoid getting into details of what measures of consciousness represent at the third level of abstraction, we can content ourselves with an understanding of their presentation at the second level of abstraction; however, once we take the step of representing consciousness as points along the axis of the cone of actualization—all that remains as a result of the objective actualization of subjective perception—we have to contend with the realization that the only way to represent subjectivity as perception of (an object), *in the same objective space as actualized consciousness,* is as a conceptualization of points along the axis of the cone of actualization of consciousness. Fortunately, when it comes to perception, we know that one point represents the object observed as it appears in the now, and the other two points represent the object as it appears and as it's observed as perception.

If it's possible to conceptualize perception of (an object) as measures situated along the axis of the cone, then what are the points representing perception of (an object) with a morphology consistent with consciousness defined by points A, B, C, and D? The conceptualization of the perception of an object, conceptualization of the inner horizon discussed in chapter two, is presented in figure 3-6 by points F, G, and H: an object observed as it appears in the now, represented by point G, is situated between the object as it's observed as perception and the object as it appears as perception, respectively represented by points F and H. (This summary is telling you figure 3-6 is presented from the perspective of the objectivity of consciousness. Perception can only be properly represented in that figure as a conceptualization.)

Sidelights

We're at a point where it might be informative to put the interpretation into perspective with a couple of interesting and important sidelights foreshadowing topics directly or indirectly addressed in the primary document. These sidelights clarify what's admittedly less than a complete picture.

For the first sidelight, you might have noticed how the encompassing of three pyramids by a circle in figure 3-2 suggests it's possible to interpret each circle in figure 3-6 as encompassing three pyramids *and their vanishing point*. Because each circle encompasses the three pyramids representing perception of (an object) and the vanishing point of those pyramids, it's tempting to interpret the geometry as suggesting consciousness of (perception of (an object)) is consciousness of the perception of both the inner and outer horizons spanning the object-observed-as-it-appears. This interpretation contradicts the hypothesis presented in chapter two claiming the perception of an object observed as it appears in the now, defining an inner horizon, collapses to consciousness. If there's validity to an interpretation where consciousness of (perception of (an object)) is consciousness of the perception of both the inner and outer horizons spanning the object-observed-as-it-appears, then we have to suspect that the figures in this chapter are communicating what happens to perception of (an object), when the object is observed as it appears, that's not communicated by the figures in chapter two.

The figures in chapter two focus on subjectivity as perception of (an object). By contrast, the figures we're presenting in chapter three focus on the relationship between objective consciousness and subjective perception of (an object), namely consciousness of (perception of (an object)), one can interpret as a focus on the transformation of perception of (an object) to consciousness. The figures in chapter three present a more comprehensive portrayal of the situation than what can be described by the figures in chapter two. If the collapse of perception of (an object), collapsing the object observed as it appears in the now of the inner horizon, turns out to be accompanied by collapse of the transcendent outer horizon of the object-observed-as-it-appears, then the collapse of both horizons is likely the result of what intentionality underlying a more comprehensive representation of the situation reveals that's not revealed by the representation of subjectivity as perception of (an object) in chapter two.[23]

The next sidelight assumes the majority of readers are at least vaguely familiar with the Nazca lines located in the southern desert of Peru. When viewed from the ground, the lines seem to have been created with no apparent rhyme or reason—scratches in the desert fading into the horizon—but our comprehension changes dramatically with a change in perspective. From the air the same lines reveal amazing geoglyphs. Perception described with respect to those geoglyphs isn't very different in methodology from the way most objects are perceived in the sense that different perspectives tend to reveal different information about an object. Most objects tend to be perceived in partiality and from one perspective at a time (Welton 1999, section 12). Perspectives can be altered to accumulate additional information about an object, yet perception of an entire object is seldom required to achieve meaningful consciousness of that object. It's an amazing ability.

If consciousness is defined as the actualization of potential represented by perception of (an object), then how is consciousness of the entire object possible when perception of an object turns out to be a partial perception represented by one perceptual act out of any number of perceptual acts? One has to suspect there's more to consciousness of (perception of (an object)) than the interpretation presented. Consciousness of an object can be achieved when consciousness of (perception of (an object)) results in consciousness of a partial perception of that object because there's another aspect to consciousness we haven't considered. Consciousness is, as argued, the actualization of potential represented by perception of (an object), but consciousness is also consciousness as subjectivity with respect to objectivity of a *thematized* object–the entire object. So it appears there is a sense in which consciousness remains undeniably consciousness of something, which implies there's more waiting to be revealed beyond our claim of consciousness as the actualization of potential represented by perception of (an object). (What's particularly fascinating is how interrelated the two aspects of consciousness will turn out to be. A complete understanding of consciousness remains elusive because of this interrelationship.)

If you're assuming what's presented up to this point couldn't incorporate details of the Great Pyramid, you're in for a surprise. As the next section will demonstrate, contributions of the Great Pyramid to the map have been with us all along. The Great Pyramid was interpreted in chapter two as representing the object as it appears as perception to help us understand perception of (an object). One should hardly be surprised if it turns out that the Great Pyramid has more to offer in the sense of knitting everything together to advance our understanding of the entire situation. Human nature is such that we tend to preoccupy ourselves with details seeming important, yet sometimes it's best to take a step back from those details and attempt to understand what's observed from the perspective of contributions to a larger narrative. Since part of the narrative has already been presented, what remains requires revealing the Great Pyramid's contributions.

The Great Pyramid and its Chambers

Figure 3-11 repeats figure 3-6 with the addition of a pyramid aligned with the axis of the cone of actualization. The base of the pyramid passes through point A, and point B suggests proximity to the location of the King's Chamber. The key observation requires recognizing the pyramid in figure 3-11 as a representation of the Great Pyramid. The angle of the face of the pyramid, the larger angle in figure 3-11 at 51.843 degrees, is the primary connection between the two pyramids. (The casing-angle of the outer casing stones, i.e. the angle of the face of the Great Pyramid, is discussed briefly in (Kingsland 1932) and (Legon 2000A) or any number of internet references.[24])

If we assume the King's Chamber is aligned with the vertical axis of the Great Pyramid, then the smaller angle in figure 3-11 at 26.565 degrees can be compared to what the angle should be, 20.442 degrees, if one intended to center point B at the King's Chamber.[25] Comparison of the angles suggests point B is located higher up the vertical axis of the pyramid than the location of the King's Chamber, but 26.565 degrees happens to be where point B should be located if it's the central point in a geometric relationship known as the Golden Pyramid. It will be convenient to refer to that central point as the Golden Pyramid point (GP). To put it simply, there's a location with respect to what's called the Golden Pyramid relationship that offers a clue to the location of point B where B = GP. (See Appendix 3 for a comparison of figure 3-11 to its realization as a Golden Pyramid.) Although you can't see it in figure 3-11, there is a connection between point B and the King's Chamber that will be revealed in figure 3-13.

We've come a long way to get to figure 3-11. Figure 3-11 will be used as the basis for the next three figures and the figures in appendix 5 to finally turn attention to an interpretation of the internal structure of the Great Pyramid. We're interested in what the internal structure of the Great Pyramid validates regarding the phenomenological situation described as consciousness of (perception of (an object)),

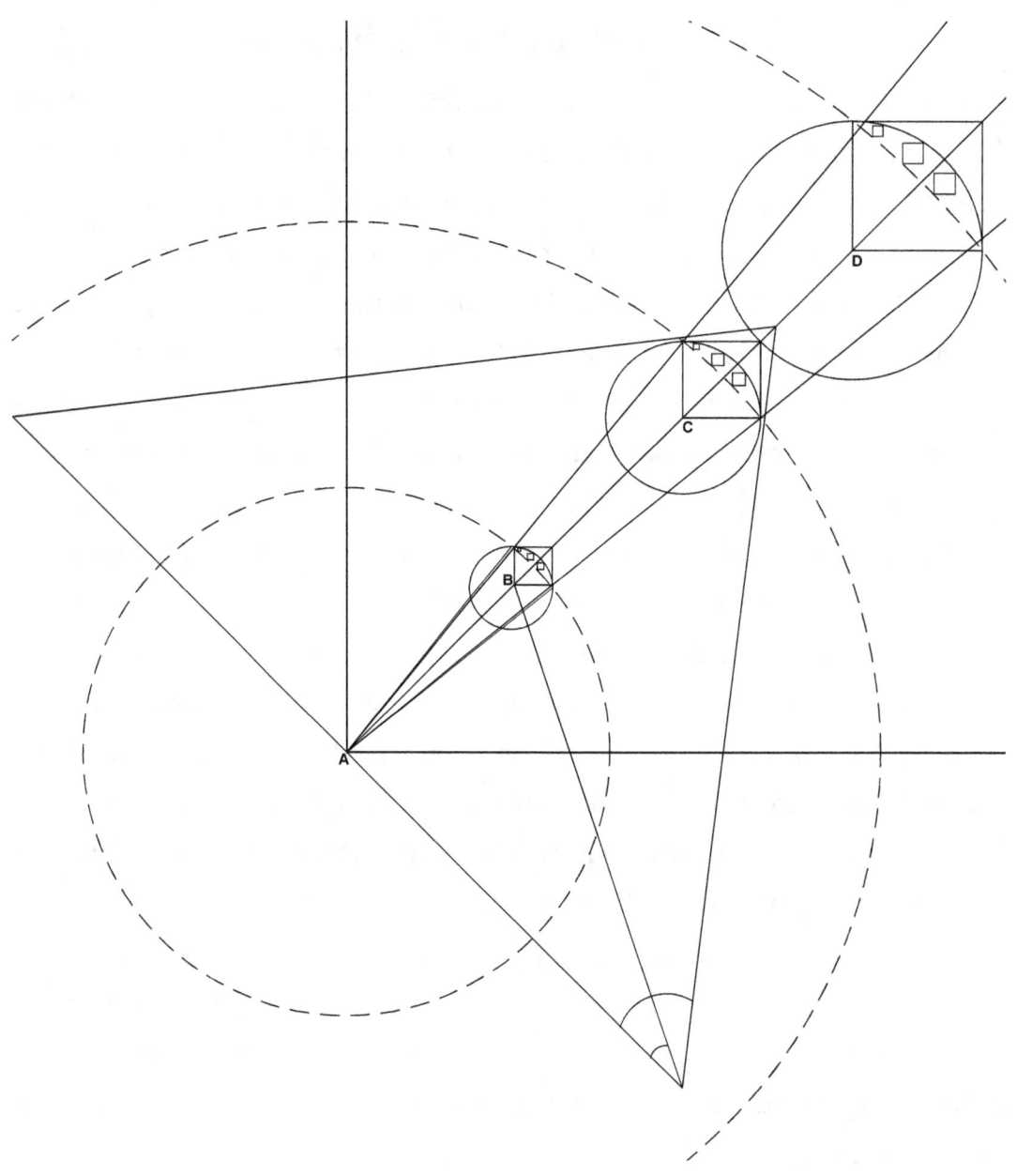

Figure 3-11

and we're also interested in any insight suggested by that internal structure beyond what's already revealed.

Figure 3-12 updates the representation of the Great Pyramid in figure 3-11 to include approximations to the locations of the King's and Queen's Chambers. What's obvious in this figure is the King's Chamber (K) isn't located along the axis of the cone of actualization of consciousness, which happens to be the axis of the pyramid, and air shafts of the Queen's chamber (Q) don't extend to the outer edges of the Great Pyramid.

Figure 3-13 introduces a subtle but significant change regarding placement of the King's chamber. Note that a dashed pyramid implied by the air shafts of the Queen's Chamber is also introduced. It shouldn't surprise us if the dashed pyramid has a story to tell. *The Queen's Chamber in figure 3-13 points to the location of point F, and if the King's Chamber is located along the axis of the cone of actualization, it points to the location of point B.* The designers of the Great Pyramid probably had good reasons for not locating the King's Chamber exactly along the pyramid's axis, but it's plausible that the configuration in figure 3-13 represents the intended communication. Let's see where this assumption can take us.

To interpret the internal structure of the Great Pyramid, the immediate focus has to be on the King's and Queen's Chambers and what they symbolize in the context of their relationship to each other and to symbolism of the overall structure. It shouldn't surprise us that the air shafts emanating from each chamber have roles to play in these symbolisms. Let's start with a general but somewhat abstract description: *each chamber points to a state at the second level of abstraction where alignments in the figure suggest each chamber represents a state of subjective certainty in related but different progressions of states.* Two air shafts from each chamber embrace a pyramid in figure 3-13 where each pyramid highlights a progression of states with a similar morphology but not the same uncertainty of observation. One might expect each state of subjective certainty to be integral to one of two progressions of states, interrelated

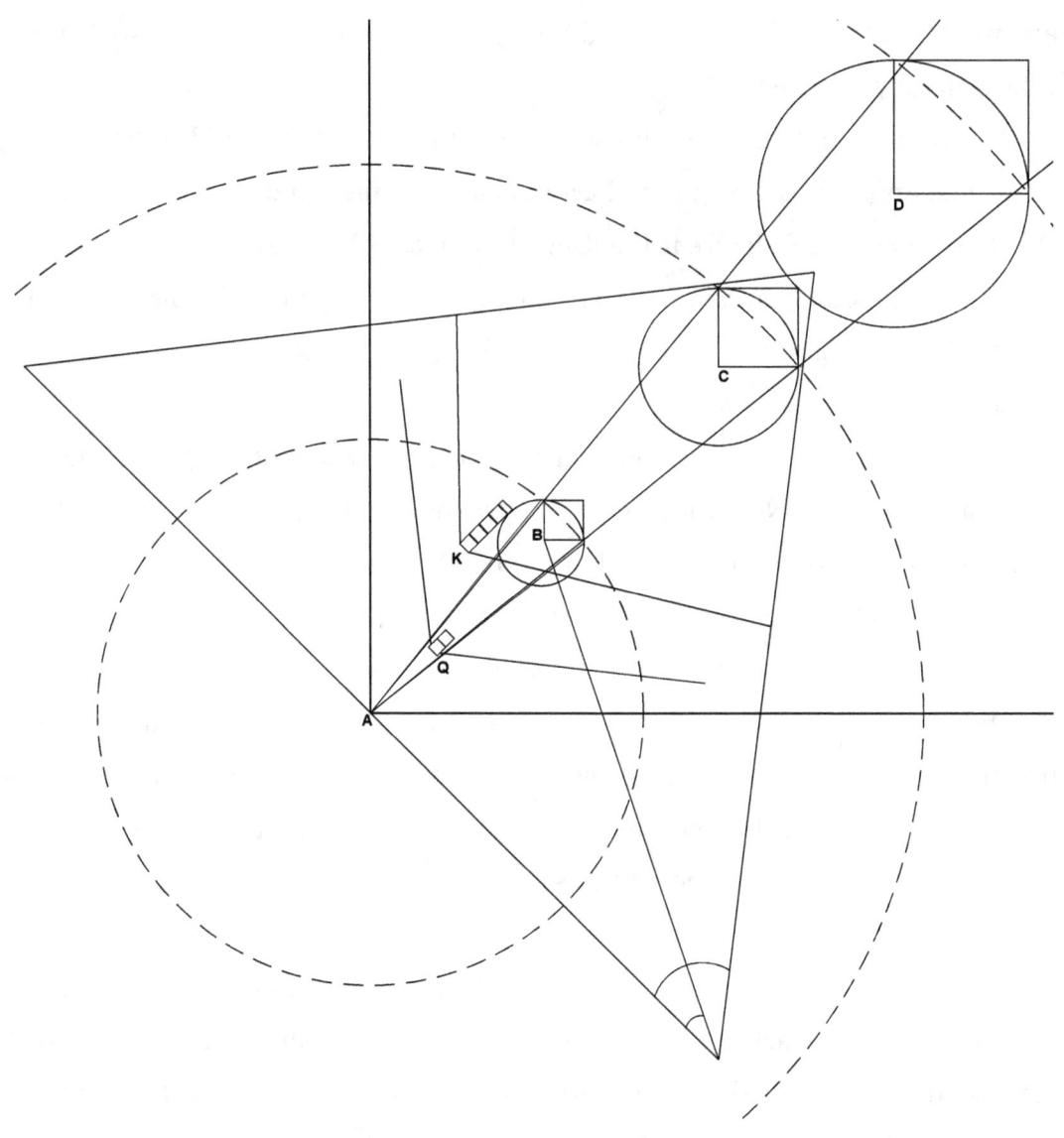

Figure 3-12

86 According to the Map

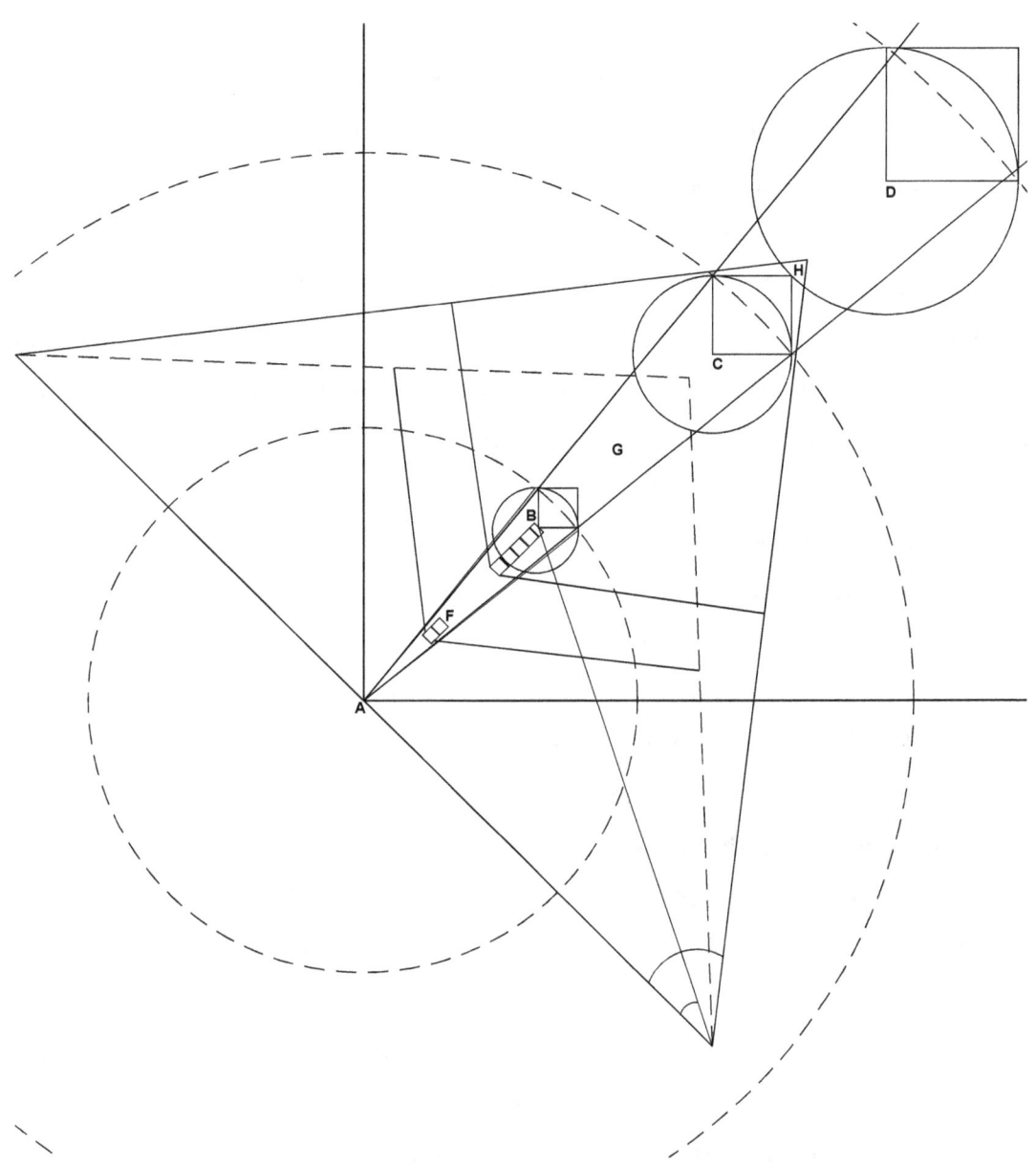

Figure 3-13

Consciousness of (Perception of (an Object))

in a manner consistent with a pyramid within a pyramid, indicative of secondary and primary contexts that come together for achievement of a common goal.[26]

To be more specific, the King's and Queen's Chambers in figure 3-13 arguably represent different states of subjective certainty: the Queen's Chamber represents a state of subjective certainty we can recognize as a subjective frame of reference at point F associated with points A, F, B, and G, where the uncertainty of observation of this pattern occurs at the state of objective uncertainty at point B; and, since the King's Chamber points to point B, it represents a state of subjective certainty we can recognize as a subjective frame of reference at point B associated with points A, B, C, and D, where the uncertainty of observation of this pattern occurs at the state of objective uncertainty at point C. (The uncertainty of observation in both cases is a distance measured from point A.) Since the air shafts emanating from the Queen's Chamber limit the size of the pyramid they embrace, that pyramid can only accommodate an uncertainty of observation at point B. The air shafts emanating from the King's Chamber, however, are capable of accommodating an uncertainty of observation at point C. (The idea that the King's Chamber represents a state of *subjective certainty* in the context of the Great Pyramid and also represents a state of *objective uncertainty* in the context of the smaller pyramid might be significant; the effect could make point B the sense of egocentricity with respect to consciousness in an *objective now* established by the relationship between two contexts.)

Something is communicated by the Great Pyramid encompassing a smaller pyramid. The sub-pattern defined by points A, F, B, and G represents the state of subjective certainty in our super-pattern. Each point in this sub-pattern (see footnote 21) is also a state of subjective certainty with respect to a sub-pattern in the super-pattern. Figure 3-13 emphasizes what we already know: the sub-pattern defined by points A, F, B, and G is a state of subjective certainty that represents the *subjective frame of reference* of the super-pattern encompassing the phenomenological apprehension of an object. Should we be surprised that the measures of consciousness and perception in the sub-pattern, points A, F, and B in particular, are temporally a posteriori with respect to the object observed as it appears in the now at point G?

Given that uncertainty increases from the measure at point A to the measure at point D, think about what the measures of consciousness and perception at points A, F, B, G, and C might have in common that the measure of consciousness at point D lacks. Why would those measures be encompassed by the Great Pyramid? In the progression from point A to point C, what is it that becomes increasingly uncertain with respect to the organism, and what does that uncertainty tell us about the measures at those points? (For more insight regarding figure 3-13, see appendix 5.)

Starting from the vanishing point relationship illustrated in figure 3-1, a relationship spanning points A and B in figure 3-6, you can't help wondering how one might acquire the knowledge needed to position additional solid squares and their encompassing circles at points C and D in figure 3-6 versus a myriad of other possible locations. Fortunately, there's an additional insight suggested by the Great Pyramid we can exploit to shed light on the choice of points A, B, C, and D that's particularly fascinating. The roof of the King's Chamber in the Great Pyramid is supported by five massive granite stones. It was obviously an intended feature, yet why it was done has always been a subject of debate.[27] What's interesting is the observation that the niche inside the Queen's Chamber also appears to represent a pyramid with five levels.[28] The designers of the Great Pyramid were trying to tell us something needs to be repeated five times; it needs to be repeated in a way that each new occurrence is built upon the one below it. What needs to be repeated is the pyramid itself, but repeat it where?

In figure 3-14, the pyramid introduced in figure 3-11 is repeated five times. The pyramid is repeated in a manner that the point representing the GP in one pyramid eventually represents the base point of another pyramid as the pyramids progress in the direction of uncertainty.[29] The pyramids progress such that every point

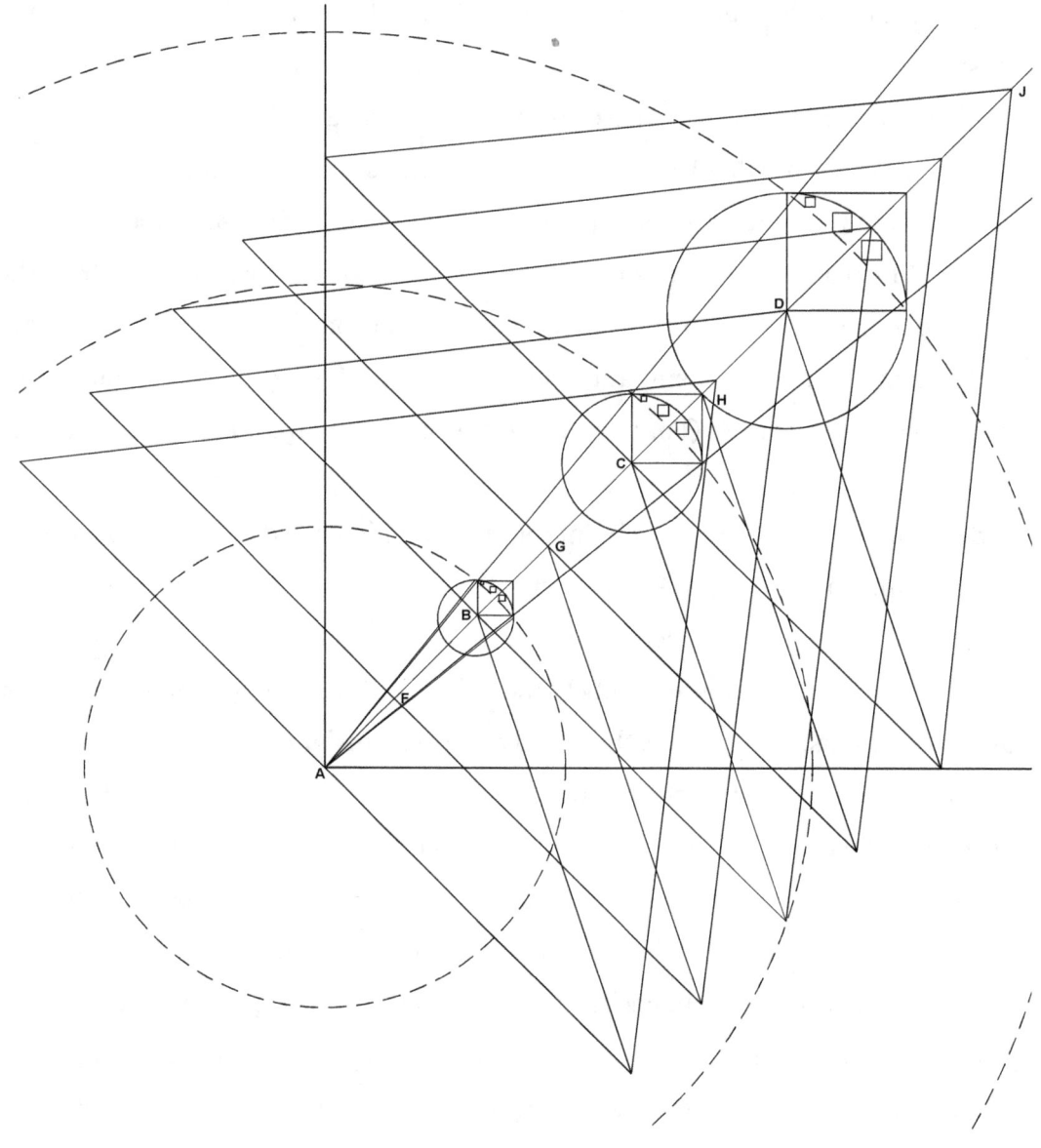

Figure 3-14

between points A and D representing a point we've associated with consciousness or perception of (an object) is either a GP or base point associated with one of the pyramids. *In other words, the GPs and base points of the progression of pyramids illustrated in figure 3-14 are the locations of all points discussed between points A and D mapping the conceptual geometry presenting the situation described as consciousness of (perception of (an object)).*

Although it's a matter of simple geometry, the alignment in figure 3-14 is remarkable. The coincidence of each point between points A and D with a GP or base point of the five pyramids projected along the axis of the cone of actualization cannot occur other than exactly where all of the points are dictated by the situation. *Associated with each point is a measure of uncertainty with respect to the certainty frame of reference at point A that isn't arbitrary at the first level of abstraction.* Coincidences of key points are consistent with the expectation that the situation of primary interest is represented by solid squares accommodating perception of (an object) at points A, B, and C, conforming to the constraint represented by the dashed square accommodating perception of (an object) at each point, and they're consistent with the expectation that perception of (an object) accommodated by the solid square at point D, not conforming to the constraint represented by its dashed square, is positioned at a measure of the uncertainty of observation beyond that of consciousness at point C. (To keep the figure as simple as possible, the dashed squares are not shown in figure 3-14. Since this figure appears somewhat busy, each step of the progression from the configuration in figure 3-11 to figure 3-14 is shown in appendix 4.)

The obvious question regarding the repetition of pyramids in figure 3-14 is why five repetitions? Why not three, four, or six repetitions? In our earlier discussion of a super-pattern encompassing four sub-patterns, we noted figure 3-6 didn't extend far enough in the direction of uncertainty to identify the sub-pattern of consciousness of (perception of (an object)) representing subjective uncertainty of the super-pattern.

The five repetitions of the pyramid in figure 3-14 extend the situation illustrated in figure 3-6 just far enough in the direction of uncertainty to represent all of the sub-patterns in the super-pattern. The required sub-pattern representing subjective uncertainty in the super-pattern is a sub-pattern supporting consciousness of (perception of (an object)) represented by points A, G, D, and J. The distance from point A to point J represents the uncertainty of measurement, the distance from point A to point G represents uncertainty of the frame of reference, and the distance from point A to point D represents the uncertainty of observation. (Once again the distance from point A to point G, uncertainty of the frame of reference, is equal to the distance from point D to point J such that uncertainty of the frame of reference plus the uncertainty of measurement equals the uncertainty of observation when uncertainty represented by the distance from point D to point J is a negative adjustment (see footnote 6 in chapter two) to the uncertainty of measurement represented by the distance from point A to point J.) What else is figure 3-14 trying to tell us?

What's Cosmic About It

The designers of the pyramids wanted to impress upon us that thinking about our own nature can't be divorced from thinking about the cosmos because the human situation is essentially an expression of the cosmos. We are the personification of an organismic universe. If the situation describing phenomenology is connected to related situations in a manner suggesting an overarching paradigm, an organismic universe, then that paradigm is cosmic in more than name if it enables a more enlightened understanding of the order of things we observe. The lack of consensus on what constitutes life is a sticking point; nevertheless, the concepts of life and organism are intimately related.

The idea that the map and its frameworks communicate a cosmic message becomes a likely proposition if the designers of the pyramids were extraterrestrial. Our primary clue that the designers were extraterrestrial is the realization that knowledge revealed by the map is considerably beyond the level of advancement of the ancient civilization we credit with building the pyramids.[30]

Suppose ancient Egyptians designed the pyramids to memorialize an interpretation of observations in the night sky, a proposition requiring less of a stretch of the imagination than extraterrestrial visitation. It's highly unlikely that an ancient civilization could have accumulated enough knowledge to recognize fundamental patterns in those observations required to formulate the generic situation presented by the map without perspectives revealed by multiple disciplines that wouldn't come into their own for at least another two thousand years, nor would they have had insight into the evolution of human civilization needed to justify encoding the conceptual geometry of the order of things—if they managed to discover it—in structures requiring a monumental investment. It can't be denied that the pyramids of Giza weren't the only pyramids built by the Egyptians, and their construction expertise had presumably advanced by the time they started building at Giza, yet there's something engaging about the pyramids suggesting the input of extraordinary intellect and wisdom when it comes to the message encoded there and how it's encoded. If ancient Egyptians were capable of such advanced insight at that point in the evolution of their civilization, it would be a real shame to think of all the knowledge they accumulated that had to have been lost over the course of history.

One could make the counter argument that ancient civilizations were more in tune with the interconnectedness of all things in the cosmos; consequently, it's hard for today's civilization to give ancient Egyptians credit for sensitivities we've only begun to revisit. Even if we allow for this possibility, there's a large gulf between symbolizing the interconnectedness of all things through the creation of deities versus acquiring the level of knowledge required to geometrically communicate the nature of our interconnectedness. Someone with in-depth knowledge of the order of things went through a great deal of trouble to gift it to us, and they did it in a way that seems nothing short of miraculous.

Summary

An argument has been presented claiming a map illustrated by the pyramids of Giza can be conceptualized as built on three progressive levels of abstraction capable of representing situations that, at the least, can be understood from the perspective of phenomenology. The phenomenological situation described as consciousness of (perception of (an object)) encompasses patterns of states that can be expanded into a framework supporting an intentional network capable of accommodating the functionalities of consciousness and perception, but what's less obvious is the idea that the framework and its measures conform to a conceptual geometry with implications for the assumed preeminence of a Cartesian subject. In other words, the map can reveal much more about what makes us the beings we are.

What makes us the beings we are is the realization that we're more than organisms interacting with our environment. We are also organisms in search of an identity. The need for identity permeates processes that would otherwise make us little more than automatons. Our inner sense of awareness as beings in a world, distinct from the world, compels us to endow consciousness with some mysterious property we hope will explain what it means to be human. It's why we cling to the idea that consciousness is more fundamental than an epiphenomenon of the brain. Although it's not obvious at the level of detail presented in the one situation we've explored, the map is telling us that our search for identity is a consequence of an order of things manifesting a conceptual geometry supporting emergent situations where the subjective cannot be easily divorced from the objective. We're in there somewhere, always just a little shy of grasping who we are.

One can't be expected to fully appreciate the contribution of the map to an understanding of the phenomenological situation without insights revealed by the third level of abstraction presented in the primary document; nevertheless, it's fair to claim we've introduced a compelling interpretation suggesting the pyramids memorialize a map encoding knowledge that could be of cosmic significance.[31] If the encoding represents the truth, then—as fantastic as it seems—we have to entertain the possibility that the designers were of extraterrestrial origin. Evidence of extraterrestrial contact not based on an interpretation would be ideal if such contact was our only interest. Fortunately, it's neither our only interest nor our main interest.

Perhaps what's more important than evidence supporting extraterrestrial contact is understanding the entire map and where it can take us, an understanding that requires resolving a number of related questions:

What cosmological model lies at the foundation of our interpretation, and how does that model support a conceptualization of the universe as an organism? How far can the organismic universe paradigm take us? What is the physical significance of the first level of abstraction upon which the second and third levels of abstraction are built? What more can be learned about temporality and time and about how both relate to the conceptual geometry represented by the first and second levels of abstraction? What is the fundamental pattern we've conceptualized as a dimension in chapter one? How can a network of interacting patterns make situations such as the phenomenological situation possible? What can consciousness as objectivity and the sense of egocentricity associated with it offer the phenomenological situation that wouldn't be possible with consciousness as subjectivity? What is the ontological interpretation of the map? What's the psychological interpretation? Can interpretation of the map provide insight into the synthesis of sensibility and understanding required for empirical knowledge? Can the map as presented support intersubjectivity? How can the map enlighten us as to the nature of the self and self-consciousness? What understanding of our place in the cosmos can one hope to achieve by following the map? Is the map communicating a teleology? Is there some deeper metaphysical message we're invited to contemplate? Finally, there's the issue of artificial consciousness.[32] If artificial consciousness is possible, then what can the frameworks tell us about how it can be achieved? All of these questions await an answer.

Chapter 4

Know Thyself

Take a Ride with Me

Have you ever been presented with an idea and wondered, "How in the world did he think of that?" This chapter will sketch the chain of events compelling me to devote considerable time and energy into research and development of the conceptual frameworks and all they reveal about the pyramids. You'd be surprised by how often the effort to understand oneself and reconcile one's place in history influences what's researched. Some of the more personal reflections have been confined to footnotes to accommodate varying levels of interest.

In 1980 I made a decision destined to change my future and open my eyes to structural economic changes with long-term impacts many of you have experienced in one form or another. That was the year I decided to move away from software development in engineering and aerospace to seek employment developing Management Information Systems (MIS). It didn't take long before I realized there was a war unfolding in corporate America, and MIS development was on the front lines of the conflict.

On one side you had the traditional Data Processing (DP) department, accustomed to developing and controlling all corporate software applications but notoriously slow to respond to development requests. On the other side were corporate departments DP supported, constantly frustrated by the lack of response to their demands for MIS. Like most wars, this one spawned technological advances. One of those technological advances was the introduction of fourth generation modeling languages. Fourth generation languages were impressive but they weren't the breakthrough dooming DP's data monopoly. That distinction belonged to the information revolution. The information revolution couldn't be stopped. New and affordable software tools provided by an explosion of desktop computers made it possible for corporate departments to free themselves of DP dependency. It didn't take long before departments were able to analyze data in ways and with a frequency previously only imagined.

From my perspective the information revolution was good news and bad news. The good news was there would be a lot more work available in the short term. The bad news was this work came at a price with respect to career development. DP departments had notoriously large budgets because the technology was expensive, employees were hired for career positions, and experienced employees expected annual pay raises. When one project was finished, you were simply reassigned to the next project on the queue. The department requesting the project was essentially your client. Corporate departments supported by the DP department, on the other hand, presented a different environment because they had considerably smaller budgets. A smaller budget meant the moment you finished developing MIS as an employee of a corporate department was the moment you became a budget liability, and, since those departments were leaner and their projects were more discretionary, you were also more vulnerable when corporations reacted to economic downturns.[1]

It became clear that as long as I developed departmental MIS applications, my average tenure with a corporation would be about three years. Unfortunately, it was an expectation that didn't account for the economic recession of the mid-1980s.[2] The recession of the mid-80s wasn't nearly as bad as the financial collapse of 2008, but it was a very serious recession affecting some states, particularly less diversified ones, more than others. States with an economy overly reliant on oil, for example, took years to recover.

I was convinced the recession was a more serious problem than those around me suggested, but there was another problem. The recession opened my eyes to trends foreshadowing long-term consequences for the economy's secondary and tertiary sectors. The pace of technological innovation and the necessity to increase productivity were destined to reshape the U.S. economy in ways the labor market seemed ill-prepared to weather: the same innovations that were going to make labor more productive were also going to make a lot of jobs obsolete—blue collar and white collar. I became preoccupied with the idea that we were on our way to structural changes in the U.S. economy that could become a serious problem if improperly managed. My biggest concern was the realization that managing change of that scale required planning for the long term in a short-term world.

If my expectations were correct, the year a worker entered the labor market in certain industries was going to determine the odds of reaching retirement before the job disappeared. Even if employees could afford to invest in some backup plan, there was no guarantee it would give them more control over their destiny in an environment where the demand for a wide range of skills would be evaporating.

It was easy enough for economists and politicians to talk about comparative advantages and becoming a service economy, but they seemed to be leaving out the reality that technology was being deployed to decrease a corporation's biggest cost: labor—that's you and me. Ironically, I was an agent of those technological changes. Call me crazy, but in the mid-80s I was of the opinion that the number of losers in the U.S. labor market in the coming decades was probably going to be huge compared to the number of winners, and, frankly, I was aware of the possibility that in such an environment I could become a casualty regardless of my skill set.[3] It doesn't matter how well qualified or experienced you claim to be if there aren't any jobs to apply for.

The unfolding economic environment wasn't just bad news for labor; it was bad news for American corporations. Increasing global competition meant a profitability squeeze was inevitable. A lot of corporations weren't going to survive in their current form: some would be acquired, some would restructure themselves, and some would move operations out of the U.S. It was too late to worry about whether those changes should or shouldn't happen. The danger was denying the reality of what was coming.[4, 5]

My assessment of the economy was a harbinger of things to come. It didn't take long before surviving the economic environment morphed from a reflection to a necessity. That necessity arose when I wrapped up an MIS project at the worst possible time. From my perspective, the worst possible time was an economy with no discretionary corporate spending, which meant no new projects and no need for specialists. I was unable to convince aerospace I'd be happy returning at the level they were hiring, and a whole generation of young engineering and computer science graduates, willing to work for even less money, was entering the market. Maybe it was time to get out of the corporate meat grinder and try something different. I did have research interests with commercial potential, and I was willing to entertain anything providing a constructive distraction from the depressing state of affairs unfolding around me.

While brainstorming about products to develop and sell, the idea occurred that in the next ten years Wall Street would become increasingly reliant on computer based models and automated trading. I was never interested in pedestrian projects, so a decision was made to engage in preliminary research on what was known in investment finance as the growth optimal portfolio (GOP).[6] The emphasis had to be on incorporating growth expectations into a market model, but the academic in me needed more than what could be extracted from financial theory. I needed a better understanding of the nature of growth, metabolism, and life cycle. To achieve that understanding, I turned to studies of organisms in an ecosystem. What ecosystem: a petri dish, a rain forest, the entire planet? What organisms? Was there a common dynamics of organismic growth, metabolism, and life cycle any ecosystem could accommodate?[7] As it turned out, enlisting the help of a number of disciplines made investigating the nature of organismic growth and metabolism more interesting than I could have imagined.

Part of the reason for choosing the GOP project was the need to find a more fulfilling environment. I couldn't afford to go back to academia, nor was I sure I'd be happy going back, so I had to find a way to bring that kind of environment to me. Perhaps research, if it could be made to pay for itself, might create an environment where I could expand my horizons. In retrospect, my actions reflected a growing awareness of the importance of environment on one's sense of self, an importance I allowed the hubris of youth to underestimate.[8]

The need for a more fulfilling environment wasn't the only reason I found research appealing. I'd been a student for so long that I clung to the notion that more knowledge would allow me to think my way out of any predicament, and, since I was typically trying to do everything, I was always painfully aware of the need for more knowledge. Acquiring knowledge for the sake of curiosity, however, would require some hard choices. Research to answer fundamental questions was probably going to be unfunded research. There were constraints, but I was free to pursue knowledge of genuine interest in whatever direction the pursuit took me.

To jump-start my research and pull myself up by my bootstraps, I decided to employ a unique approach to creative thinking. For lack of a better word, I'll describe the approach as Synthesizing.[9] I coupled Synthesizing with a technique called The Focusing, a name borrowed from my research. The Focusing is nomenclature to articulate a focusing of knowledge. Initially the strategy appeared to be working. By Synthesizing around a unifying concept, I was able to accommodate a great deal of information from seemingly unrelated disciplines; and by giving my imagination free reign while guiding it with the information I was accommodating, in the way I was accommodating it, I was converging on insights about the order of things I didn't think anyone, including myself, understood. When The Focusing was working, I could literally feel energy surging through my fingers. Nothing's more magical than the unfolding of the truth.

If I could have stepped out of myself, I would have noticed something was changing. I allowed my research to take on a worrying personal element by imposing early theories on certain members of the academic community. Just the thought that certain individuals might read the material was enough motivation to keep pumping it out. If ever there was a bad idea it was that one. First, imposing your theories or point of view on others is totally unprofessional. Second, early investigations of common patterns unifying what most people consider disparate disciplines weren't well developed or completely understood. However, each time a working paper was mailed out I put a little more pressure on myself to figure it out.[10] I was becoming a driven man. My ego was beginning to overrule reason in scary ways, and, to make matters worse, I was documenting it. I was completely out of control. How did I let that happen? What set of circumstances put me in the position of acting against my better judgement?

Have you ever seen one of those science fiction movies where some poor soul, under control of divine or supernatural forces, starts spewing out all kinds of numerology with some deep prophetic meaning no one can figure out? In retrospect, I claim to have experienced similar but less dramatic events. I wasn't spewing numbers out; I was recognizing them in all kinds of phenomena subject to spatiotemporal uncertainty. Some hidden order was revealing itself ever so subtly. There was no divine or supernatural influence, but there was a price to be paid. I had to employ more dimensions of my mind without losing focus or losing my mind in the process. I had to keep Synthesizing and Focusing without letting the technique consume me.

There had to be a way to break through the barriers holding me back—a way to keep assimilating more and more information. Something had to give.

It's the ultimate irony: I was beholding glimpses of the truth about the order of things, but I couldn't see the truth unfolding about myself. It took quite a few years before I fully grasped I was going through what can be described as a *crisis of self*. As the lifestyle I allowed to define me disintegrated, my sense of self began to disintegrate. I reacted by turning my research into an attempt to prove to the world and to myself that I mattered. (I'd been in various degrees of a conflict of self for most of my adult life. What happened was a disheartening combination of events and revelations, at that point in my life, turned a conflict of self into a crisis.[11])

We all carry a certain amount of baggage with us through life. For me, it was a conflict of self rooted in the self's ephemerality. A conflict of self isn't necessarily a bad thing, nor is a conflict of self limited to individuals. (At the time of this writing, it can be argued the United States and a number of Western European countries are all showing symptoms of identity crises caused by changes in their environments. Large groups in each country are interpreting their environment as changing in ways making them uneasy about whom they are or the extent to which they matter.[12]) Things we do to resolve our conflict of self provide pivotal moments in human development that can lead to profound insight with the potential of altering the course of history. Unfortunately, what can be self-defeating and sometimes dangerous are the things we do when we don't know what to do or when anxiety becomes more than we can handle. Misguided efforts to cope with our conflict of self can rob us of the opportunity to sustain or reinvent ourselves in a changing environment.

To resolve what was becoming a downward spiral, bootstrapping my mental machinery via Synthesizing and Focusing had to be abandoned. I miss the energy surges but there were some scary side effects: some of the nightmares were rather frightening, and the stress was beginning to affect my health. Sometimes when you let the genie out of the bottle you don't get three wishes, and it can be really hard to get him back in again. The most expedient way to stabilize the situation was to reinvent myself, yet for all that was gained by reinvention something was lost–in America we call it losing your mojo.[13]

So what happened to the GOP? It was only a matter of time before the project had to be shelved due to a lack of funding. I wasn't overly disappointed. The GOP wasn't a project of significant emotional investment; it was a means to a financial end. On the downside, the real cost of the project was the substantial lead time and energy required to undertake the effort.[14] The GOP research did serve a vital purpose. To make progress in the research, I needed my mental wheels to start turning again, but, to keep them turning, I needed more than the GOP could provide. While attempting to reconcile the GOP research with other disciplines, it became clear that there could be more to cosmology than theoretical physics and astronomy. Perhaps I needed an understanding of the cosmos that was situated between physics and metaphysics but still fundamental and robust enough to incorporate other disciplines. To that end, an organismic interpretation of the cosmos became appealing.

You might think I left the corporate world to indulge in research and never looked back, but that wasn't possible. Over the years, I've periodically put my research on the back burner to accommodate the need to make a living.[15, 16] I even tried to walk away from it a few times only to find it invariably beckoned my return. Research provides a sense of purpose in a world where so many appear to lack a sense of purpose. It satisfies the desire to understand the world and one's self. When engrossed in research I'm thinking, "Maybe this is where I'm supposed to be."

The research I'm involved in these days definitely benefits from insights gained from those early days of Synthesizing. Am I still Synthesizing? If I am Synthesizing, the technique is tempered by successive refinement and lessons learned. I don't experiment with The Focusing anymore. The Focusing was fueled by a level of information overload successive refinement avoids, and it may be a state of mind requiring the vitality of youth. It served its purpose.

How did I end up thinking about the pyramids? One of the journals of interest as an undergraduate would occasionally present a geometric proof without words.[17] In a geometric proof without words, the entire statement of the proof is contained within the geometry; all that's required is enough insight to interpret what's presented as a proof. My visit to the pyramids as an undergraduate found me wondering if the structures might represent a geometric proof worthy of the effort. There was something haunting about the geometric simplicity of the pyramids that demanded explanation. If the proof turned out to be a map, I can live with that.

I'll close by recalling that my first encounter with a computer was in the early seventies when it occupied an entire room and required a staff to keep it operational. Communicating with the behemoth required a big deck of cards and a paper printout. It may be hard to imagine but in those days interacting with a computer from a programmer's perspective always had the air of something special. The primary investment was, as it remains today, some function of time–time to design the program, time to debug it, time for it to execute, turnaround time, time to figure out what happened–time that could have been used to learn or do something else.

One project required writing a program to play a board game against the computer.[18] I remember opening the printout to discover the computer congratulated my program's success in defeating its program. That kind of feedback was appealing because it animated a perceived opponent. It also validated decision making contingent upon the occurrence of a state of nature among possible states of nature as situations on a board game my program anticipated and took advantage of to realize its goal. Decisions were executing in a short period of time where the game could be played as many times as funds would allow. The computer's program could be modified to learn from experience, and I could modify my program to learn from experience.

Once you made it beyond the initial trauma of how to program the thing, it was fascinating to learn how a computer actually worked. Understanding what a computer couldn't do made understanding what you could do seem even more miraculous. The experience changed my life.

Is it possible my research in artificial consciousness represents an attempt at recreating a bygone romance with a machine now more common than a toaster? Perhaps. Why is artificial consciousness a rewarding pursuit? Isn't it obvious? Consider the presentation you're reading and the implication of extraterrestrial life in the universe more sophisticated than humanity. Think about all we need to learn about the mind in order to bring artificial consciousness to fruition. Frameworks presented in the primary document, concepts we've barely touched upon, represent a bridge artificial consciousness must cross.[19] The gratification isn't immediate but it doesn't have to be. The journey is the reward. I'm inviting you to take this journey with me.

References

Alison, Jim. 2008. "Proposed Design Specifications for the Great Pyramid and for the Siteplan of the Giza Plateau." http://home.hiwaay.net/~jalison/Art3.html.

Baghramian, Maria. 1998. "Why Conceptual Schemes?" In *Proceedings of the Aristotelian Society*. New Series, Vol. 98, 287-306. Wiley.

Bauval, Robert and Adian Gilbert. 1994. *The Orion Mystery* . Three Rivers Press.

Bertalanffy, Ludwig Von. 1968. *General System Theory*. George Braziller.

Burge, Tyler. 2010. *Origins of Objectivity*. Oxford University Press.

Cook, Robin. 2002. "The Case For A Plan Centred On Khafre." Originally located at www.sevenislands.50webs.com/the_horizon_of_khufu/plan.htm. (Site and content recently revised (see analysis.htm); status of plan.htm is unknown.)

Devitt, Michael and Kim Sterelny, *Language & Reality: An Introduction the Philosophy of Language*. 1987. MIT Press.

Dokic, Jerome and Elisabeth Pocherie. 2006. "On the Very Idea of a Frame of Reference." In *Space in Language: Linguistic Systems and Cognitive Categories*, edited by M. Hickmann and S. Robert, 259-280. John Benjamins.

Dreyfus, Hubert L., and Harrison Hall, editors. 1982. *Husserl Intentionality & Cognitive Science*. The Massachusetts Institute of Technology.

Drummond, John J. 2003. "The Structure of Intentionality." In *The New Husserl,* edited by D. Welton, 65–92. Indiana University Press.

Eddington, Arthur S. 1949. *Fundamental Theory*. Cambridge University Press.

Føllesdal, Dagfinn. 1982. "Brentano and Husserl on Intentional Objects and Perception." In *Husserl, Intentionality, and Cognitive Science,* edited by Hubert L. Dreyfus, 31–41. Bradford Books.

Frie, Roger. 1997. *Subjectivity and Intersubjectivity in Modern Philosophy and*

Psychoanalysis: a study of Sartre, Binswanger, Lacan, and Habermas. Rowman and Littlefield Publishers, Inc.

Goodfellow, Stephen. 1987. "Giza Pyramids; The Vanishing Point." www.goodfelloweb.com/giza/index.html.

Gorner, Paul. 2007. *Heidegger's Being and Time An Introduction.* Cambridge University Press.

Guignon, Charles B. 1983. *Heidegger and the Problem of Knowledge.* Hackett Publishing Company.

Held, Klaus. 2003. "Husserl's Phenomenological Method." In *The New Husserl: a critical reader,* edited by Donn Welton, 3-31. Indiana University Press.

Horst, Steven W. 2011. *Symbols, Computation and Intentionality: A Critique of the Computational Theory of Mind.* CreateSpace, Charleston, SC, USA.

Kingsland, William. 1932. *The Great Pyramid in Fact and in Theory.* London: Rider and Company.

Klatzky, Roberta. 1998. "Allocentric and Egocentric Spatial Representation: Definitions, Distinctions, and Interconnections." In *Spatial Cognition – An Interdisciplinary Approach to Representation and Processing of Spatial Knowledge (Lecture Notes in Artificial Intelligence 1404),* edited by C. Freska, C. Habel, and K. F. Wender, 1–17. Berlin: Springer-Verlag.

Kodiak, Sterling. 2014. *The Framework of the Mind: what's required for artificial consciousness.* First Draft (Version 23). (Contents subject to revision and title subject to modification.)

Lauer, Quentin. 1978. *The Triumph of Subjectivity: An Introduction to Transcendental Phenomenology.* Fordham University Press.

Legon, John. 2000A. "The Design of the Great Pyramid." In *Egyptology and the Giza Pyramids.* www.legon.demon.co.uk/greatpyr.htm.

Legon, John. 2000B. "The Plan of the Giza Pyramids." In *Egyptology and the Giza Pyramids.* www.legon.demon.co.uk/gizaplan.htm.

Legon, John. 2005. "Vanishing Point Update – September 2005." www.legon.demon.co.uk/vpupdate.htm.

Levison, Stephen C. 2004. *Space in Language and Cognition: Explorations in Cognitive Diversity.* Cambridge University Press.

Lohmar, Dieter. 2003. "Husserl's Type and Kant's Schemata." In *The New Husserl: a critical reader*, edited by Donn Welton, 93-124. Indiana University Press.

Lyons, William. 1995. *Approaches to Intentionality.* Oxford University Press.

Massey, Heath. 2015. *The Origin of Time: Heidegger and Bergson.* SUNY Press.

McIntyre, Ronald. 1982. "Intending and Referring." In *Husserl Intentionality & Cognitive Science*, edited by Hubert Dreyfus and Harrison Hall, 215–231. The Massachusetts Institute of Technology.

McIntyre, Ronald and David Woodruff Smith. 1982. *Husserl and Intentionality.* D. Reidel Publishing.

McIntyre, Ronald and David Woodruff Smith. 1989. "Theory of Intentionality." In *Husserl's Phenomenology: A Textbook*, edited by J. N. Mohanty and William R. McKenna, 147–179. Washington, DC: Center for Advanced Research in Phenomenology and University Press of America. (Tends to be available on the internet.)

Meilinger, Tobia and Gottfried Vosgerau. 2010. "Putting Egocentric and Allocentric into Perspective." In *Spatial Cognition VII, LNAI 6222*, edited by C. Holscher et al., 207–221. Berlin: Springer-Verlag.

Mesle, C. Robert. 2008. *Process-Relational Philosophy: an introduction to Alfred North Whitehead.* Templeton Press.

Moran, Dermot. 2000. *Introduction to Phenomenology.* Routledge.

Muller, Richard A. 2016. *The Physics of Time.* W. W. Norton & Company.

Musto, Alexandra. 1999. "On Spatial Reference Frames in Qualitative Motion Representation." Technical Report FKI-230-99. Institute für Informatik, Technische Universität München. Also available at https://www7.in.tum.de/um/research/groups/ai/fki-berichte/postscript/fki-230-99.pdf.

Neuhouser, Frederick. 1990. *Fichte's Theory of Subjectivity*. Cambridge University Press.

Organski, A.F.K. and Jacek Kugler. 1980. *The War Ledger*. University of Chicago Press.

Pietersma, Henry. 2000. *Phenomenological Epistemology*. Oxford University Press.

Rescher, Nicholas. 1996. *Process Metaphysics: an introduction to process philosophy*. SUNY Press.

Schneider, Eric D. and Dorion Sagan. 2005. *Into the Cool: energy flow, thermodynamics, and life*. University of Chicago Press.

Schrödinger, Erwin. 1967. *What Is Life?* Cambridge University Press.

Searle, John R. 1983. *Intentionality: An essay in the philosophy of mind*. Cambridge University Press.

Sherover, Charles M. 2003. *Are We in Time?: and other essays on time and temporality*. Edited with a preface by Gregory R. Johnson. Northwestern University Press.

Smith, Barry, and David Woodruff Smith, editors. 1995. *The Cambridge Companion to Husserl*. Cambridge University Press.

Sokolowski, Robert. 2000. *Introduction to Phenomenology*. Cambridge University Press.

Torrai, Gábor, and George Kampis, editors. 2005. *Intentionality Past and Future*. Rodopi B. V., Amsterdam – New York.

Welton, Donn, editor. 1999. *The Essential Husserl: Basic Writings in Transcendental Phenomenology*. Indiana University Press.

Widdows, Dominic. 2004. *Geometry and Meaning*. CSLI Publications.

Yau, Shing-Tung and Steve Nadis. 2010. *The Shape of Inner Space: String Theory and the Geometry of the Universe's Hidden Dimensions*. New York: Basic Books.

Zahavi, Dan. 2003. *Husserl's Phenomenology*. Stanford University Press.

Appendix 1

Figure A-1 can be found in Robin Cook's "The Case For A Plan Centred On Khafre."

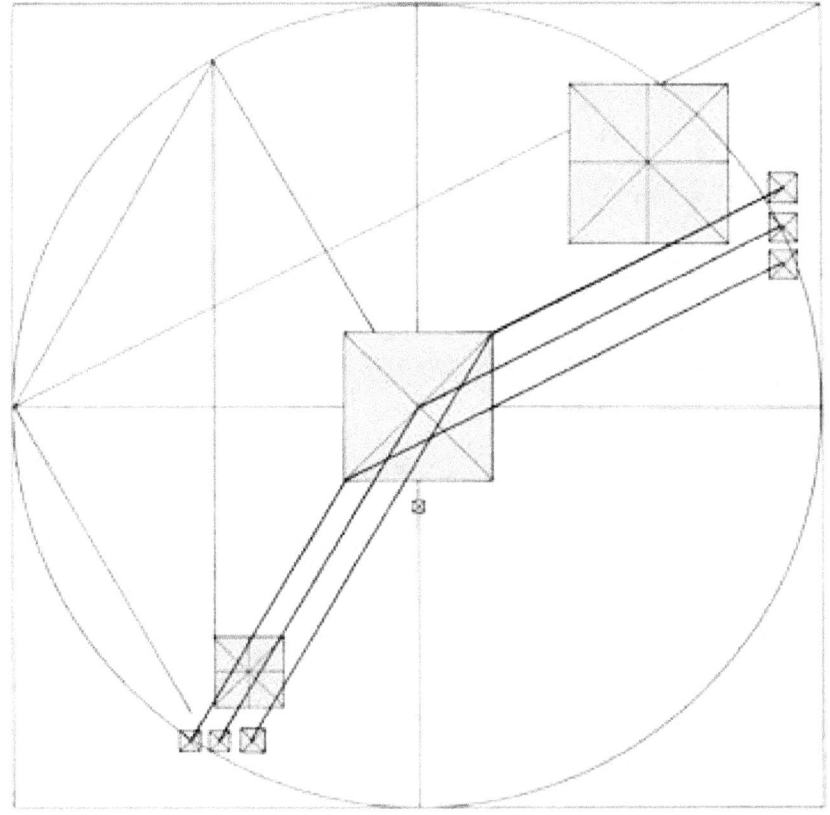

Figure A-1

111

Appendix 2

Figures A-2 and A-3 illustrate Legon's original (1987) analysis of the Giza Vanishing Point. Discussion of these figures can be found at www.legon.demon.co.uk/vpupdate.htm. The figures can also be found as part of a letter from Legon to Goodfellow on Stephen Goodfellow's web site at www.goodfelloweb.com/giza/041687.html.

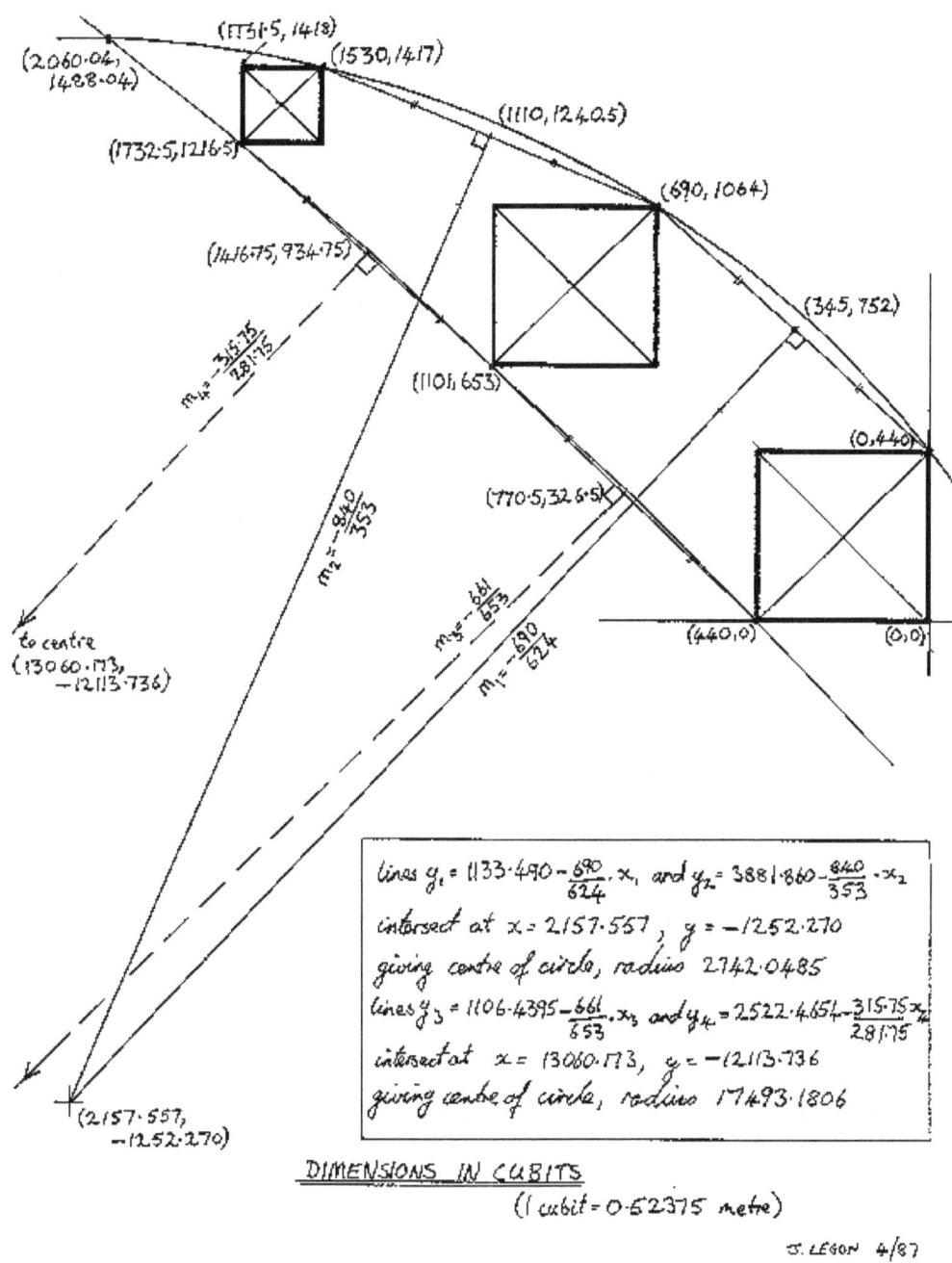

Figure A-2

Appendix 2

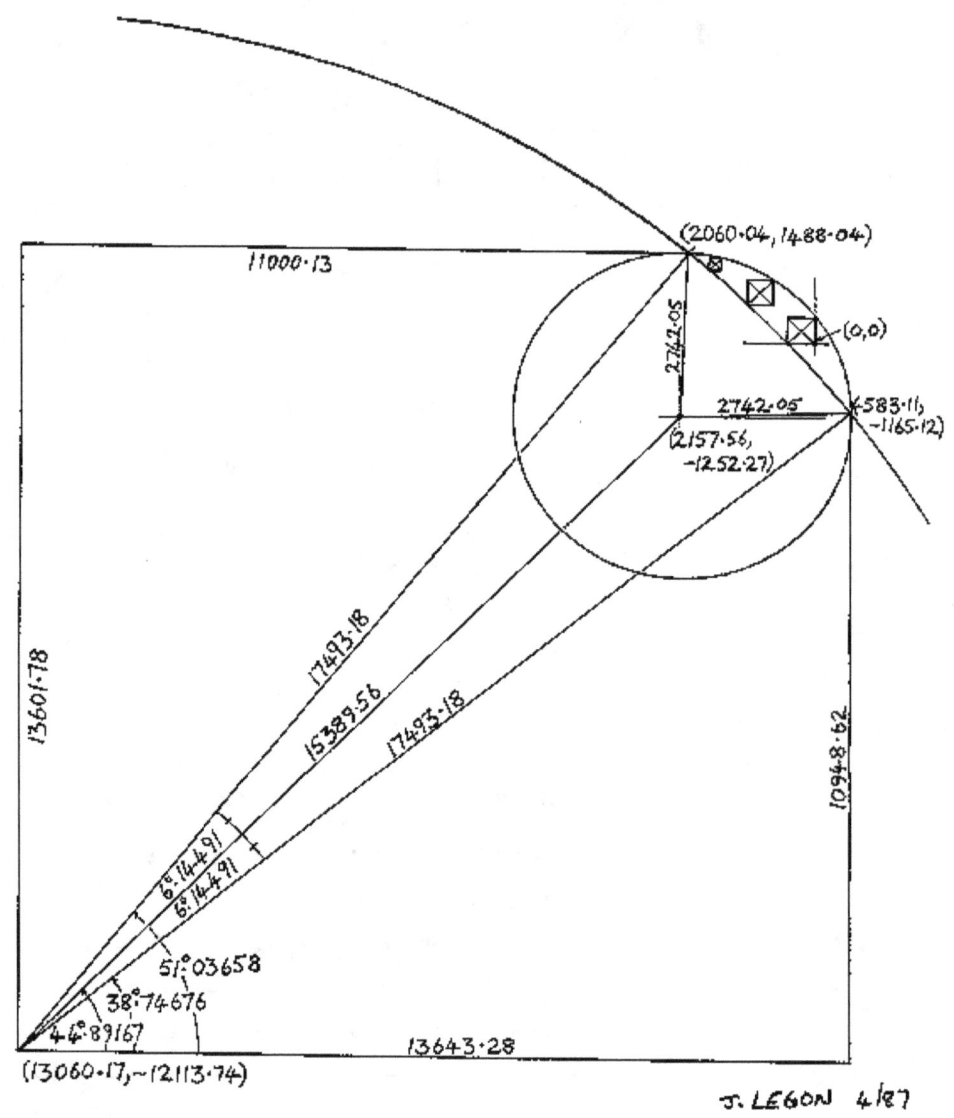

Figure A-3

Appendix 3

In figure A-4, we've extended the height of the pyramid to reflect what it would be if it conformed to a Golden Pyramid geometry where the line connecting points A and X is scaled such that AX = 2.61803 = (1.61803)^2 = (phi)^2. Phi is known as the golden ratio. This adjustment would increase the face angle of the Great Pyramid from 51.843 degrees to 52.6225 degrees. (Numerous references and diagrams exploring the relationship between the golden ratio and the Great Pyramid can be found on the internet.)

Why wasn't the Great Pyramid constructed as a Golden Pyramid? Examination of figure A-4 suggests it's reasonable to assume the physical dimensions of the Great Pyramid were chosen for their consistency with consciousness of (perception of (an object)) illustrated between points A and H. The Golden Pyramid relationship simply isn't as good a fit with the relationships illustrated within the cone of actualization as is the Great Pyramid. The important distinction between the Golden Pyramid and the Great Pyramid in this case is the exact value of the face angle. It was the intent of the designers of the Great Pyramid to communicate a face angle of 51.843 degrees. We won't go into the details of why that measure is important other than noting it's a communication pertinent to the conceptual geometry at the first level of abstraction.

Take another look at figure 3-8. The non-coincidence of point O with point B is such that line segments determining the cone of actualization (see figure 3-7) aren't wide enough to pass through points N and L suggested by figures 3-1 and 3-2. (You might recall that the vanishing point relationship within the circle centered at point O in figure 3-2—a relationship described by the pyramids at Giza—is the vanishing point relationship depicted at point B in figures 3-7 and 3-8.) The circle at point O not being coincident with point B is one of those nagging discrepancies that can really haunt you, particularly when you have a scheme where everything else seems to be falling into place.

What if the cone of actualization was wider to the extent that its sides did pass through points N and L? Line segments determining the cone of actualization drawn from point A, through points N and L, are shown in figures A-5 and A-6 such that they intersect the larger pyramid representing the golden pyramid relationship at points Y and Z. A circle intersecting points Y and Z can be centered along the cone of actualization, but, as shown in figure A-6, it's centered just beyond point C in the direction of uncertainty. If there's meaningful information in figure A-6 to help us understand the discrepancy illustrated in figure 3-8, then it's possible the information is communicated by some measure associated with the circle passing through points Y and Z, which happens to be larger than the circle centered at point C associated with the original cone of actualization shown in figure A-4. Information communicated by the circle passing through points Y and Z, if it reveals anything, would probably reveal some type of limit.

We won't go into the details about the limit this circle might communicate because it would require a discussion at the first level of abstraction; however, even though it's hard to believe anyone could be so clever, it would be an ingenious way of communicating a value that might otherwise be impossible to communicate via the message as it's geometrically presented. Is it possible they counted on us trying to resolve the discrepancy? If as we progress we fail to explain the discrepancy illustrated in figure 3-8, we should come back to figure A-5 and reconsider whether any fundamental constant or limit the figure might communicate would be worth introducing a discrepancy to bring it to our attention. We simply haven't presented enough details about the message, particularly at the first level of abstraction, to engage in such reflection at this time.

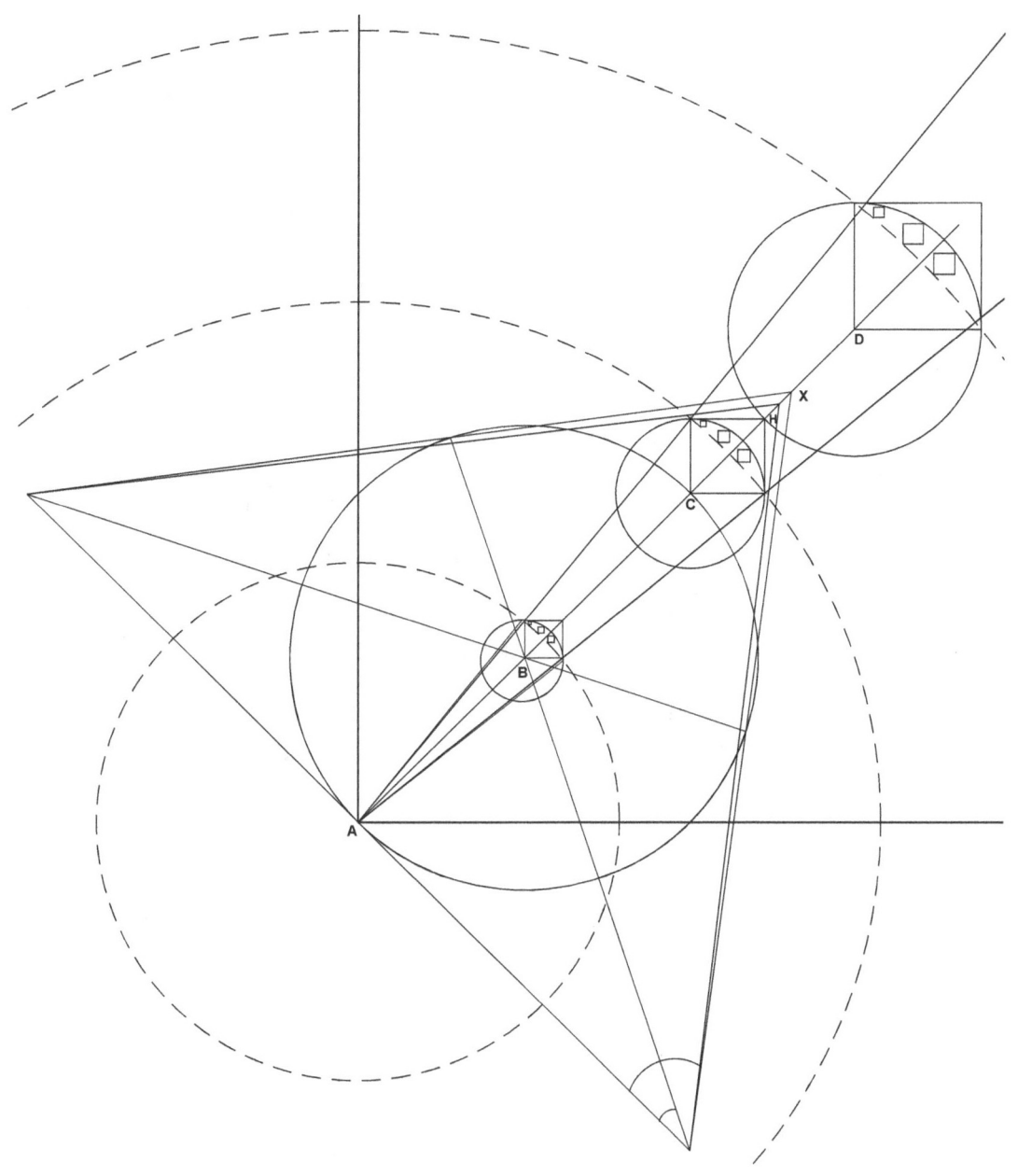

Figure A-4

Appendix 3

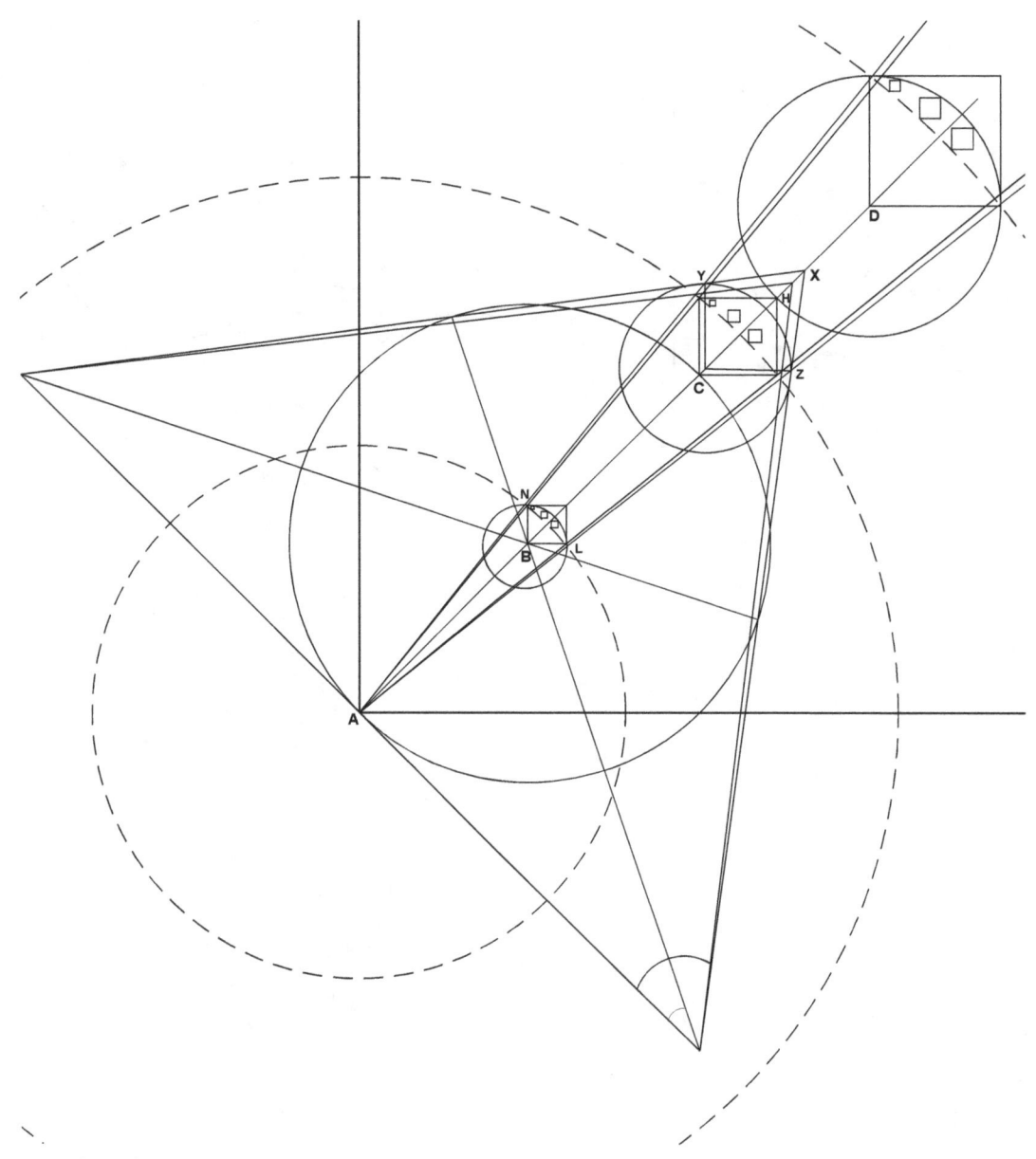

Figure A-5

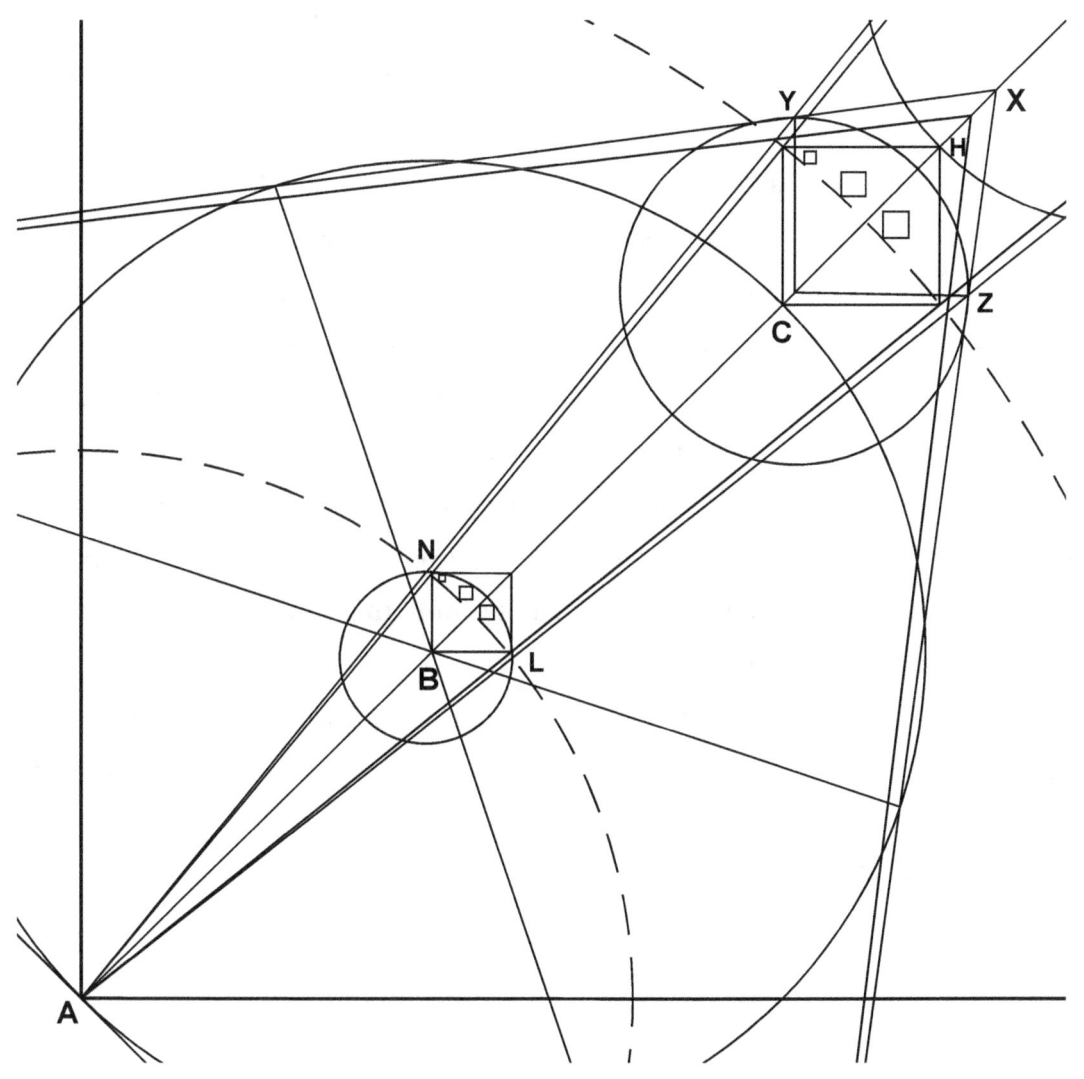

Figure A-6

Appendix 3

Until we demonstrate otherwise, what we've suggested with the aid of figures A-5 and A-6 must be considered highly speculative—a less abrasive way of saying it could easily be wrong. Nevertheless, it's an interpretation with a certain appeal to the wonder of it all: an example typical of speculations, associations, or ideas presenting themselves along the way. Most of the time our speculations turn out to be wishful thinking or coincidences, but every now and then some insight of value is identified. The trick is to judiciously document the most promising speculations, assuming they fit the context of the presentation, so they're not lost over time as we progress in our own research or lost in the passing of the baton from one researcher to another. In this sense, the presentation you're reading is produced for your benefit and ours.

The map encoded in the pyramids at Giza isn't speculation; it's likely the accumulation of a mature body of knowledge the pyramids' designers were intimately aware of. Layers of abstraction, boundary conditions, and exact values purposely encoded in the design of the pyramids are important; they imply a level of sophistication and knowledge that couldn't have been consistent with respect to the overall development of an ancient civilization unless we take into account the possibility of outside influence. There's knowledge encoded in the pyramids ancient Egyptians should not have known, yet now that we understand it—at least a portion of it—the evidence demands an explanation.

Appendix 4

Figures A-7 through A-11 are provided to show the progressive layering of five pyramids from the configuration shown in figure 3-11 to that of figure 3-14 where figure A-7 is figure 3-11 and figure A-11 is figure 3-14. Keep in mind that the dashed squares introduced in figures 3-5 and 3-6 are not shown in an attempt to keep the diagrams as simple as possible, but the constraint upon the situation imposed by those dashed squares remains valid.

Figure A-7 is essentially the same as figure 3-11 except for the inclusion of points F, G, H, and J. Points F, G, and H represent the conceptualization of perception of (an object) along the axis of the cone of actualization where, because of representation of the situation described as consciousness of (perception of (an object)) along that axis, these points (as well as points A, B, C, and D) can't occur anywhere along the line connecting point A to point J except exactly where you see them. *The position of each point along the axis is dictated by an underlying network of intentionality conforming to geometry capable of accommodating a multidimensional conceptualization.* So, without knowledge provided by the lower levels of abstraction, what would give you any clue to the locations of all the points along the axis of the cone of actualization representing the situation defined between points A and D?

The pyramid in figure A-7 only gives us a clue to the location of points A and B, but watch what happens as we progressively introduce the next four levels of pyramids in figures A-8 through A-11. All of the points mentioned above intersect a line segment associated with at least one of the pyramids. It's not a coincidence; it's a clue provided by particulars of the Great Pyramid. Do you really think an ancient civilization had that kind of knowledge about the cosmos?

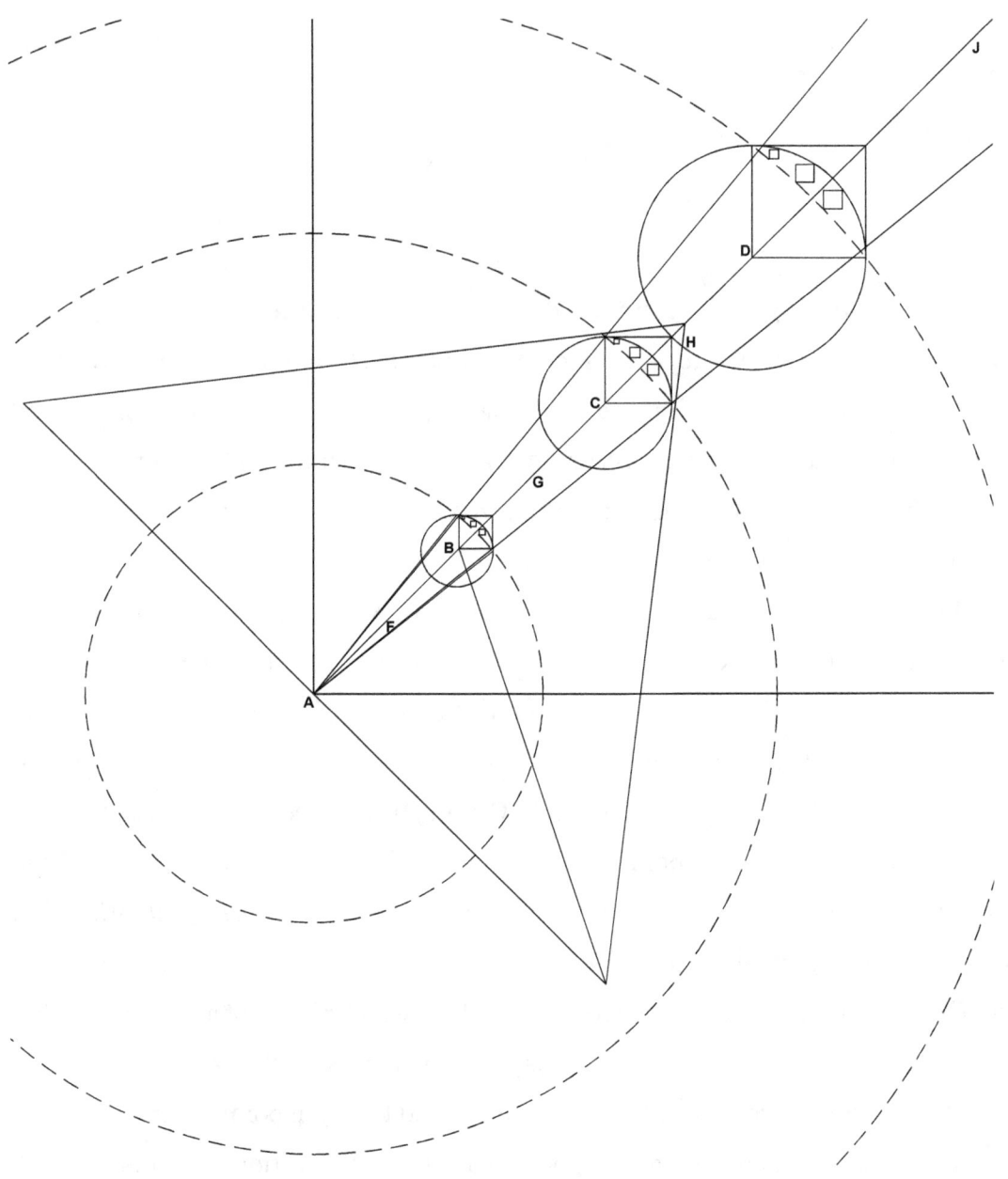

Figure A-7

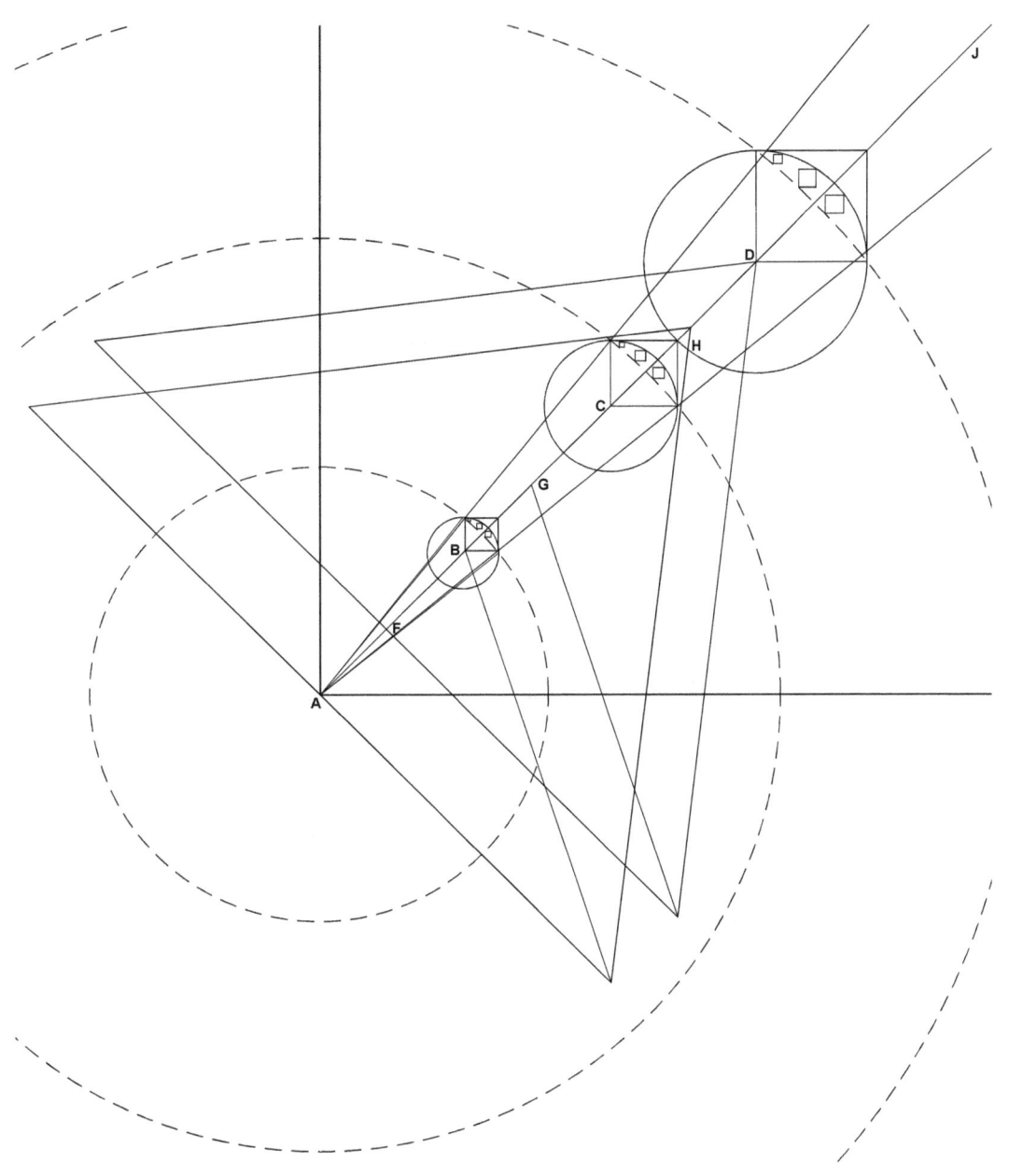

Figure A-8

Appendix 4

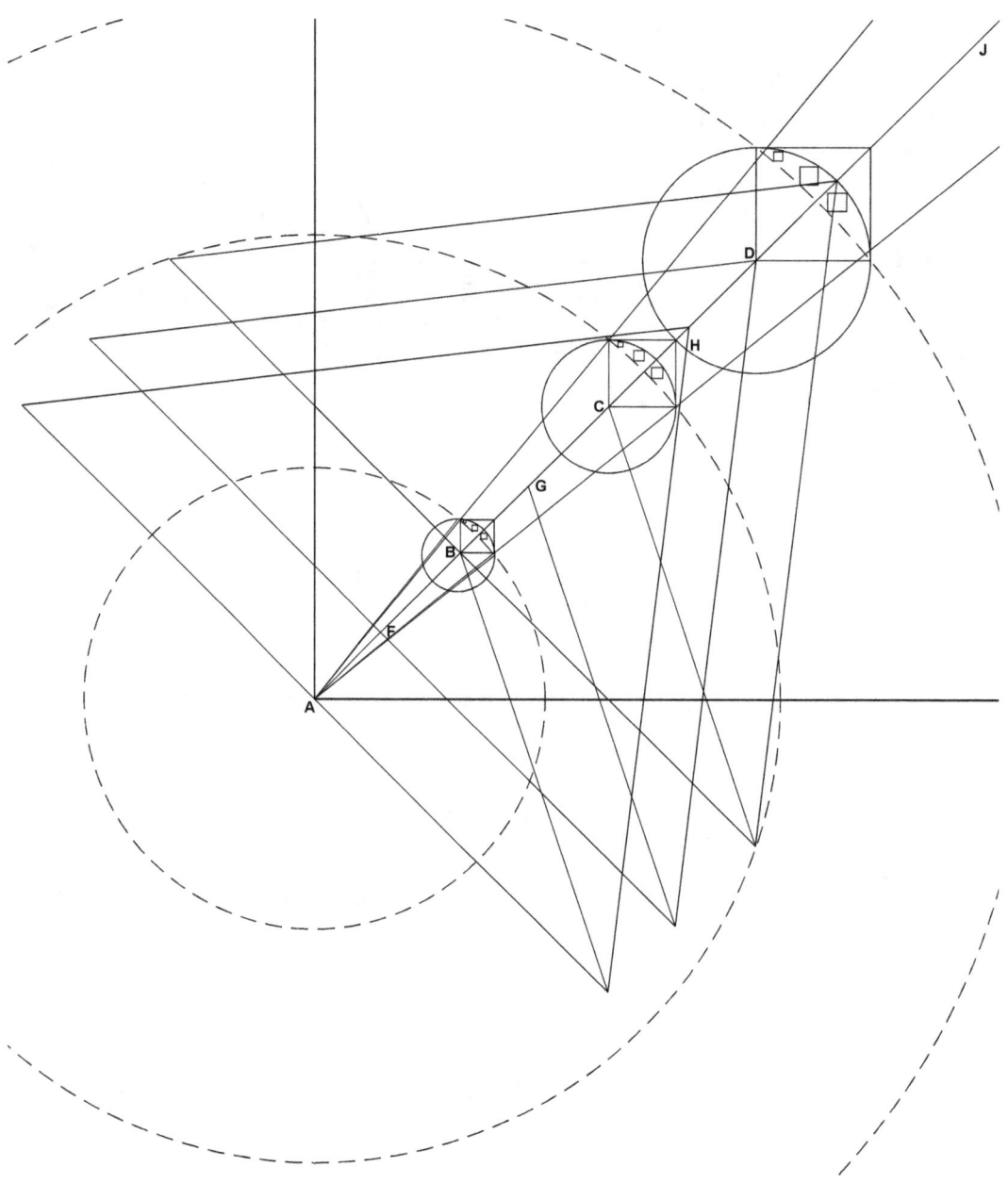

Figure A-9

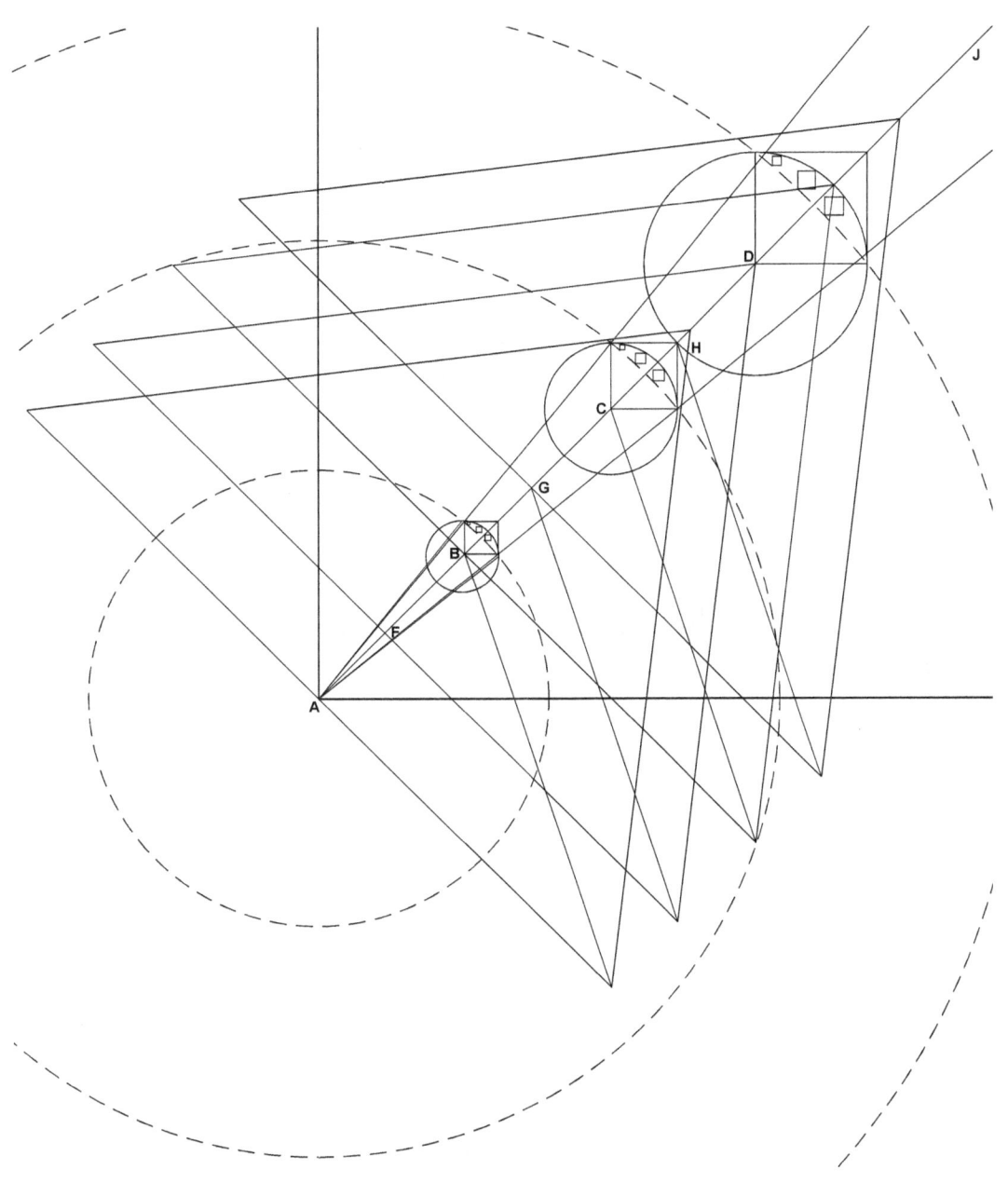

Figure A-10

Appendix 4

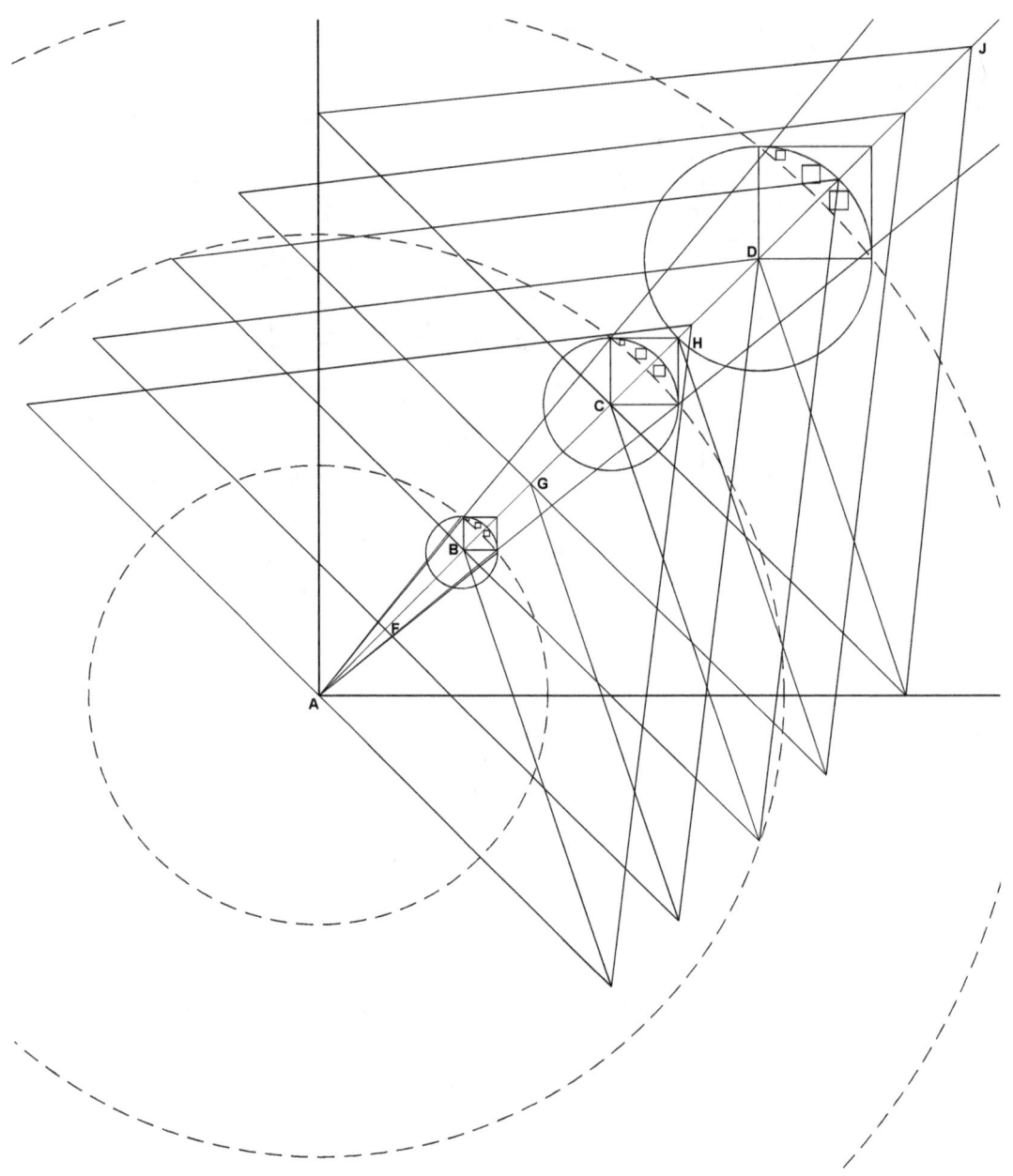

Figure A-11

Appendix 5

In figures A-12 and A-13, we've repeated figure 3-13 with the intent of interpreting what might normally be dismissed as a curiosity. Take a look at the circle centered at point B in figure A-12 and note that it's bounded by the smaller pyramid suggested by the Queen's Chamber and its air shafts. That circle encompasses points F, B, and G. Now note that a circle of the same radius in figure A-13, encompassing points B, G, and C, is framed by the Great Pyramid. Unfortunately, the circle isn't centered exactly at point G, but it's undeniably close. Are these two circles nothing more than geometric coincidences or is there information to be gleaned from their relationship?

The circle centered at point B in figure A-12 is centered on a measure we want to associate with the objectivity of consciousness, and the circle centered at point G in figure A-13 is centered on a measure we want to associate with the subjectivity of perception of (an object). Based on what we currently know, the most interesting measure of consciousness in the sub-pattern defined by points A, B, C, and D occurs at point B, a state of subjective certainty that's the subjective frame of reference of consciousness; and the most interesting measure of perception of (an object) is the object observed as it appears in the now at point G.

We have to assume it's not a coincidence that the circle in figure A-13 presents a measure of subjective perception (at point G) in some relationship with two measures of objective consciousness, nor is it a coincidence that the circle in figure A-12 presents a measure of objective consciousness (at point B) in some relationship with two measures of subjective perception. If the relationships in these circles aren't coincidences, then what are they telling us?

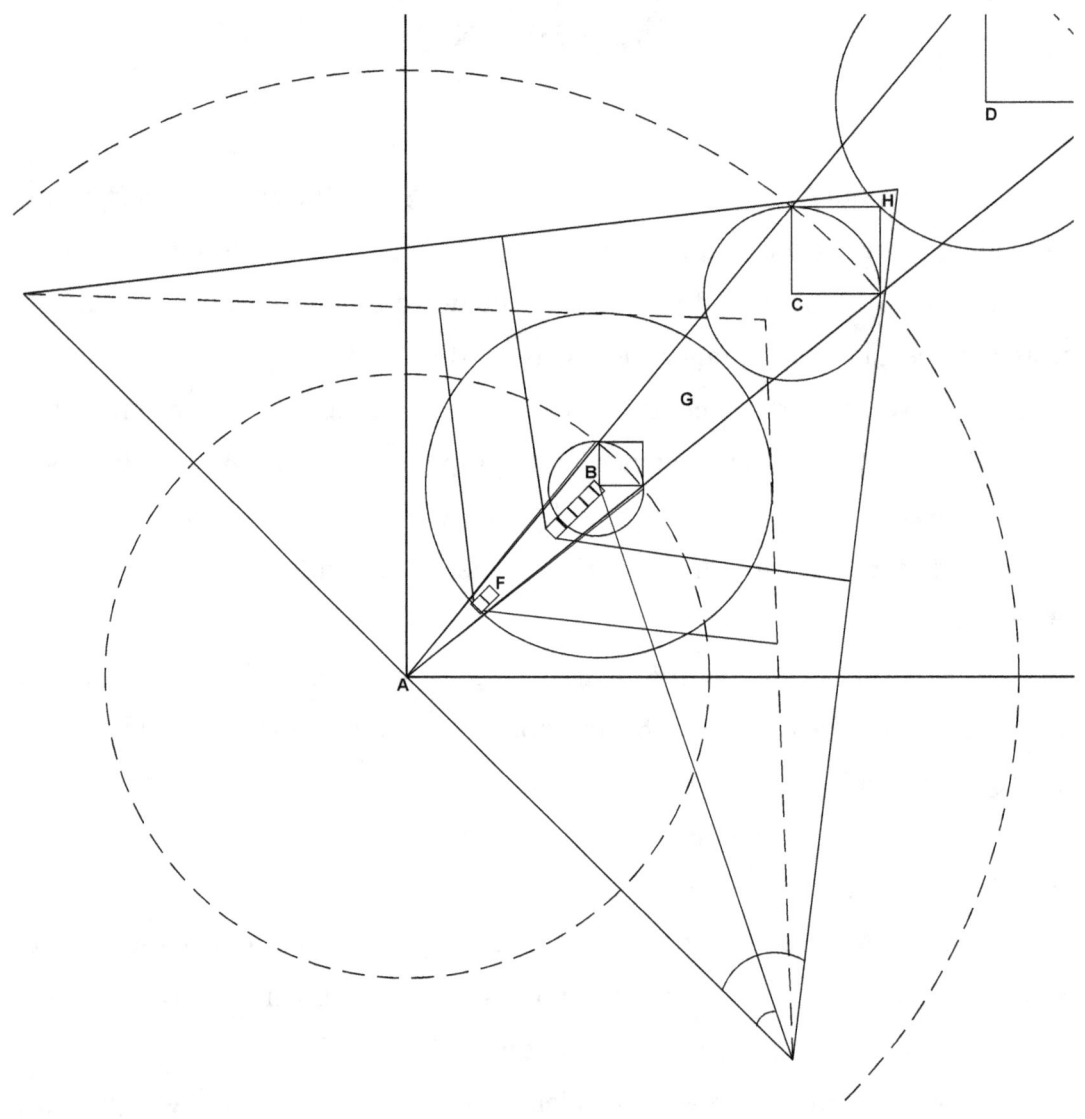

Figure A-12

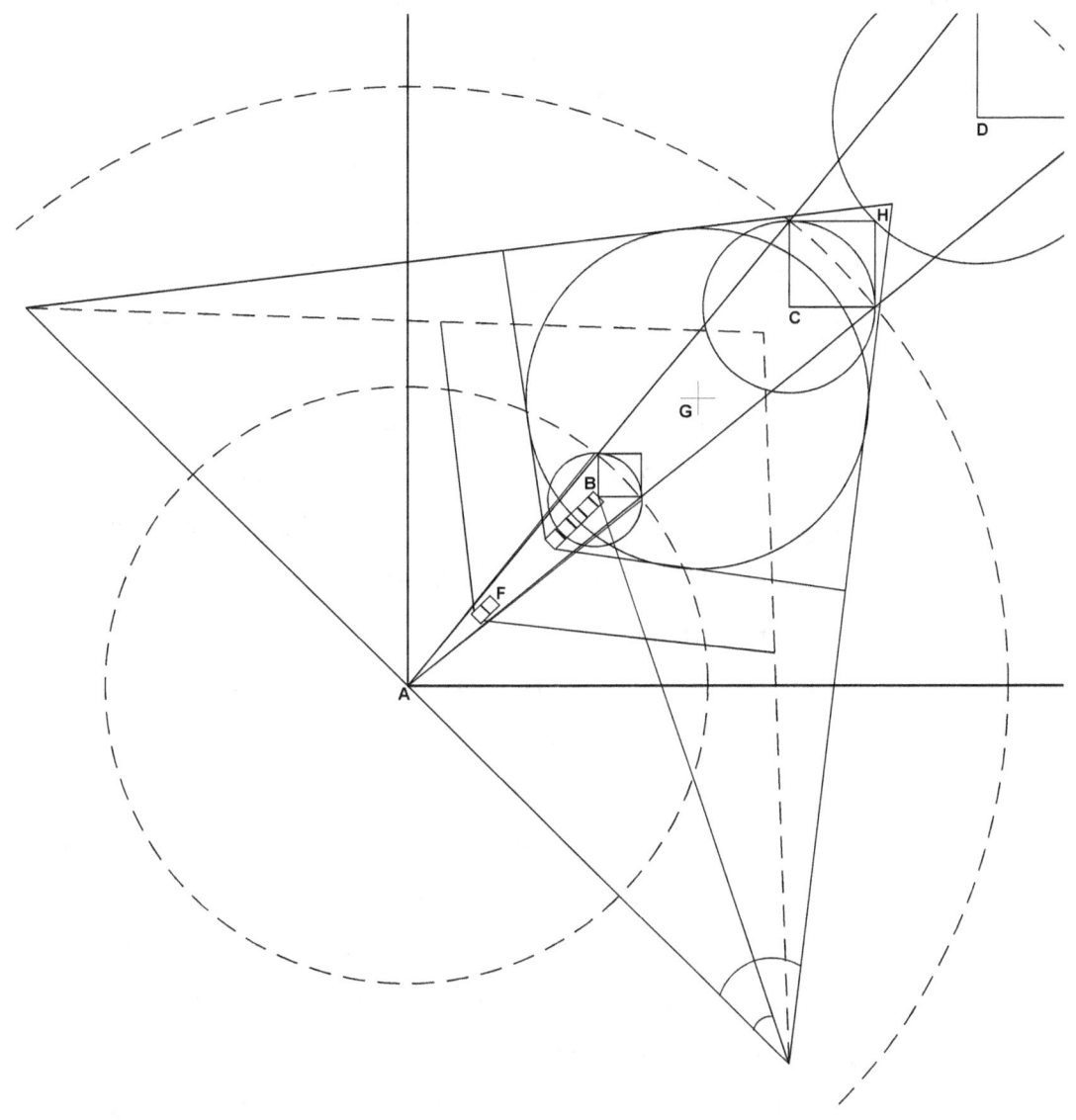

Figure A-13

Appendix 5

Figure A-13 may be communicating information we're now in a better position to understand. The Great Pyramid encompasses points A, B, and C of the four points representing objective consciousness, but it doesn't encompass point D. The measure of consciousness at point A has a relationship with the measure of consciousness at point C because the former is a measure of objective certainty and the latter is a measure of objective uncertainty. Yet we now know that the measure of consciousness at point C, according to figure A-13, is likely to have some relationship with the measures at points B and G. The measure of consciousness at point D, a measure of subjective uncertainty, has neither of these relationships.

Note that both circles have two measures in common: a measure of objective consciousness at point B and a measure of subjective perception at point G. Given the observation that the relative position of each measure within each circle changes, it's reasonable to assume that each figure is telling us something different about the measure at the center of its circle.

One of the possibilities we might consider is the idea floated in chapter three that the measure of consciousness at point B, a measure that gives objective consciousness its sense of egocentricity, represents what can be described as an objective now with respect to the subjective now at point G. Even though the measures of perception of (an object), presented along the axis of the cone of actualization of objective consciousness, are conceptualizations with respect to the measures of objective consciousness, the figures show that with respect to a temporal progression from uncertainty to certainty, an objective now at point B would be more certain than a subjective now at point G. Is there anything else an objective now might offer consciousness that's not provided by a subjective now?

Finally, note that point H in Figure 3-13 represents the object as it appears as perception, but we can't be sure that the measure would have been represented by the capstone of the Great Pyramid. Point H might, however, explain why the Great Pyramid has no capstone. Look at its location with respect to the apex of the pyramid.

Notes

Preface

1. One could argue that the generic situation mapped by the pyramids is a conceptual framework. It's an argument where a specific situation becomes an interpretation of that framework. Even though treating the generic situation as a conceptual framework has the appeal of one framework accommodating multiple interpretations, we run into problems when it comes to identifying the related concepts the conceptual framework incorporates. To be useful, each interpretation of the conceptual framework must be a conceptual framework.

2. You shouldn't be surprised that a message encoded in one of the most awe inspiring structures in the history of mankind, eluding decoding after centuries of speculation, might not be trivial. Use the diagrams, take your time, and let this presentation be a stepping stone to further research and reflection.

3. See (Kodiak 2014). We currently have no plans to release our research to the public beyond what you're reading, but plans have a way of changing.

4. The exercise of putting this document together has revealed additional insights necessitating revisions and expansion of the primary document. Since the focus of the primary document is development of integrated frameworks that enable situations we're claiming are consistent with a conceptual geometry illustrated by the pyramids at Giza, we may be inclined to introduce material making both documents less independent.

chap 1 fn 1-4

Chapter 1

1. Apprehension in this case is best described as phenomenological apprehension contributing to experience of an object. Phenomena of interest will be objects claimed to be observed as they appear, apprehended via subjective perception, prior to expression as objective consciousness.

2. Intentionality is the argument introduced by Franz Brentano (Moran 2000, chap. 1) that all mental acts, particularly acts of consciousness, are directed toward an object. It would be up to Husserl (Welton 1999, 86–112) to develop an interpretation capable of addressing the shortcomings of intentionality proposed by Brentano. For more on intentionality with respect to Husserl, see (McIntyre and Smith 1989; Føllesdale 1982; McIntyre 1982; Drummond 2003). A number of informative contributions regarding the relationship between intentionality and Husserl's noema can be found in (Dreyfus and Hall 1982). For related references, see (Torrai and Kampis 2005; Searle 1983; Lyons 1995; Horst 2011). *Intentionality and concepts that expand intentionality's utility, such as Husserl's noema and epoche, are lumped together in this presentation by describing all of it as a network of intentionality.*

3. What's not discussed in this presentation is the idea that the mind plays a unique role in the phenomenological apprehension of an object that enables an object to be observed as it appears in the now. Although it's hard to discuss intentionality, perception, or consciousness without assuming a mind, we don't explicitly discuss the mind because we're trying to avoid assumptions about the *subject*. In the situation we're presenting, the intentional network–not the mind–plays a unique role in the phenomenological apprehension of an object as it progresses from an object as it appears to an object as it's observed. Without insights provided by all of the frameworks, what we think of as a subject with a mind needs to remain a tacit assumption. (See footnote 4.)

4. There are two types of situations: those we're in and those we observe. It might be instructive to take a step back for a moment and consider what happens when we observe a situation and attempt to understand it. When we recognize a situation and choose to relate entities in that situation by assigning each one a value or location in some multidimensional space, we're essentially associating each entity with an appropriately chosen value of a metric. Should we choose to concern ourselves with how one entity relates to another with respect to understanding a situation, then that comparison arguably requires a conceptual geometry. If, for example, you step into an elevator and observe five females ranging in ages from a few weeks to over ninety years old–an incident that actually happened–and all of them appear to be related, you can hypothesize that the situation represents five generations of the same family by creating *or recognizing* a conceptual geometry of a family tree (even if it's represented by a simple progression in a multi-dimensional space) that enables you to assign each female a position in the tree indicative of her relative age. (For more on geometry and meaning see (Widdows 2004).)

Throughout the presentation we'll make reference to a conceptual geometry and conceptual frameworks built on that geometry, but what is it exactly a conceptual framework has to offer? Let's start by relating a conceptual framework to an example of what it allows us to develop that should be easy to appreciate: If you're using any kind of software on a computer or cell phone where the interface is graphic or a windowing system, then the application handling each tap on the screen or click of the mouse is processing that event using an *application framework* specifically designed to process an event on that particular platform. An application framework makes a collection of software resources available to an application developer to relieve the developer of the burden of writing lower level software to take care of

chap 1 cont fn 4

details regarding the look and feel of the application's interface and flow of control dictated by the handling of each event. (One of the primary features of an application framework is the framework calls low level code in software libraries to perform lower level tasks, compared to the traditional approach where the same code is called by the application. This feature represents a design principle described as *inversion of control*.)

Is an application framework a conceptual framework? There are many application frameworks that can make life easier for software engineers, but what do these frameworks have in common? For the sake of conversation, let's separate applications into two classes: web applications and desktop applications. We can reasonably describe a generic application framework for each class of application. Each generic application framework will be inherently conceptual and consequently qualify as a conceptual framework. *We can describe a conceptual framework as a structure connecting a collection of concepts in some meaningful way to make their interrelationships easier to comprehend from a convenient perspective. Implicit in the framework is the encapsulation of details of each concept.*

In this presentation, we won't go into details regarding conceptual frameworks at the third level of abstraction; however, we will present enough for you to notice that what might be described elsewhere as a state, the value of a property of an object, or a concept might be referred to here as a *measure*. A measure is terminology with the appeal of being generic enough to accommodate the discussion of different levels of abstraction without abandoning dependence on context for a complete understanding. (The history behind our use of the term measure stems from a decision to avoid describing a concept as the value of a variable in a generic description described by equations. There is a segment of the target market with members whose eyes glaze and ears begin to fold the moment they see or hear frameworks described with equations and variables. Conceptual frameworks are presented diagrammatically as relationships between symbolic representations. These representations can subsequently be described in a more detailed manner that's consistent with the objectives and comfort level of the target audience.) Use of the measure terminology will also be convenient in describing representations and their functionalities without constantly referring to them as representations—something we're trying to avoid. The downside is use of the terminology adds one more term like representation, entity, state, attribute, value, or concept one needs to reconcile with existing literature.

The interpretation we're offering assumes the conceptual frameworks mapped by the pyramids present concepts related to each other as patterns that can be understood from an external perspective as subject-object relationships. What's recognized as the subject and what's recognized as the object depend on what we choose to represent. It's not unreasonable, however, to suspect patterns in a framework supporting a phenomenological situation could come together such that some of their subject-object relationships are consistent with *phenomenological epistemology*, a consistency that would require the phenomenological framework accommodated by the map to be the kind of conceptual framework discussed by (Piertersma 2000, 21-26) with respect to phenomenological epistemology. The conceptual framework discussed by Piertersma is considered conceptual because the subject-object relationship defining the framework is internal: there are no objects in the framework unknowable to an internal subject. (For additional comments about an internal subject, see footnote 3 in this chapter.) In this case the conceptual framework has an internal aspect allowing a user of the framework to use it as an instrument to gain *cognitive access* to entities. (The idea that a conceptual framework consistent with phenomenological epistemology—also

chap 1 fn 5-6

referred to as a conceptual scheme—is mapped by the pyramids needs to be approached cautiously. The relationship between conceptual schemes and reality is not without controversy. For more background on conceptual schemes, see (Lohmar 2003, 93-124; Devitt and Sterelny 1987, part IV; Burge 2010, chap. 6; Baghramian 1998).)

5. Robin Cook (Cook 2002) used the same perspective of the Giza complex to present the relationship as a curiosity. His observation was mentioned as part of a presentation suggesting the plan of the Giza complex might be centered on the middle pyramid known to Egyptologists as Khafre. It was presented as a curiosity, but it would have been reasonable to suspect the observed relationship was intended by the pyramids' designers as communication of information relevant to Khafre. Cook's figure has been repeated by permission of the author in Appendix 1. (See Cook's website for his latest thoughts.)

6. Details of the fundamental pattern represented by each pyramid in figure 1-3 and of how interaction between fundamental patterns realizes a conceptual geometry supporting capabilities we associate with a cognitive organism are discussed in the primary document. The suggested multidimensionality, by the way, is not speculation. The beauty of the map is it's founded on multidimensionality of the order of things discovered long before recognizing a connection with the pyramids. Without that multidimensionality neither the frameworks in the primary document nor discoveries revealed here would be possible. (Implications of a multidimensional order of things are still being explored, which makes this all very exciting—and why we haven't released more details. Who could have encoded knowledge of a multidimensional cosmos into the design of the pyramids?)

The idea that two primary pyramids in figure 1-3 can be combined with the nine subsidiary pyramids in the same figure, suggesting the possibility of an eleven dimensional interpretation of the order of things, is timely, yet you have to wonder if there's additional information of significance in the size differential between the primary and secondary pyramids.

7. What is phenomenology? Reconciling consciousness of (perception of (an object)) with a first person perspective is offered as a simplistic description of phenomenology; this isn't how it's typically defined. Phenomenology as method and philosophy came into its own with the work of Edmund Husserl. For a range of references defining phenomenology, particularly with respect to phenomenology espoused by Husserl, see (Lauer 1978, chap. 1; Piertersma 2000, chap. 2; Sokolowski 2000; Smith and Smith 1995; Zahavi 2003). (Held 2003) is also recommended.

Think of the above description in the context of a related reflection: if knowledge of an object is rendered through a situation described as consciousness of (perception of (an object)), how can we be sure objects we perceive as existing in a world, objects of which we claim consciousness, reflect reality? What reveals itself in the situation described as consciousness of (perception of (an object)) might be interpreted as apprehension of an object as phenomenon in contrast to knowing that object as noumenon. In other words, if what separates us as cognitive organisms from noumena is the situation we've just described, then if we have knowledge of noumena that knowledge must be in the form of essential knowledge provided by phenomena.

8. In using the phrase consciousness of (perception of (an object)), it's obvious that the assumed relationship between a perceiving organism and the perceived object is not reciprocal. Our perception and consciousness of an object seem intuitive, but it's not intuitive that an inanimate object might have perception or consciousness of us. How do we account for the realization that as beings in a world we necessarily interact with that world—we act upon it and it acts upon us? Some type of feedback mechanism between us

and the world around us must be at play. At what level of abstraction does this feedback manifest itself? It's a tacit assumption of the situation that we know ourselves as the organism to which perception and consciousness belong. What makes this assumption valid? Where is our subjective frame of reference? We'll have more to say about the phrase consciousness of (perception of (an object)) at the end of chapter two.

9. For more background on string theory you might start with (Yau and Nadis 2010).

10. Additional patterns we could claim are fundamental will be introduced supporting consciousness of (perception of (an object)), but they'll be patterns built on the multidimensionality discussed here. In other words, we have fundamental patterns conceptualized as multidimensional and additional fundamental patterns that multidimensionality enables.

Chapter 2

1. We're going to attempt to enhance the flow of this presentation by referring to what is subjective as *subjectivity* and by referring to what is objective as *objectivity*. (Although a lot of repetition in the presentation could have been avoided, the primary motivation for the repetition is clarity. We want the reader to spend less time trying to understand what's communicated so more time can be devoted to exploring implications of the communication.)

2. The perception of an object has a subject-object relationship with consciousness where consciousness is, in some sense, equivalent to that perception, provided the object is observed as it appears. Describing consciousness of the perception of an object as the actualization of a potential is a convenient analogy that will allow us to visualize how consciousness and perception are interconnected, but, without formalization, it's only an analogy.

It's important to appreciate that the collapse of perception of (an object) is demanded by a network of intentionality designed or evolved to accommodate an object observed as it appears in the now. The collapse of perception of (an object) is not enabled by consciousness; the collapse is enabled by the dynamics inherent in the network of intentionality the situation connecting consciousness and perception is built upon.

3. In one of the sidelights in chapter three, we'll suggest there's still a sense in which consciousness remains consciousness of something.

4. Since details about the first level of abstraction will not be presented, this might be a good time for a few miscellaneous comments.

The pyramids represent a generic situation. In chapter three, an interpretation of the generic situation as a phenomenological situation will be subject to a boundary condition consistent with constraints at the first level of abstraction. The boundary will reveal itself at point C in figure 3-6. (Point D is also the expression of a boundary condition but it will not be discussed.) The designers of the map presented by figure 3-6 could not have designed it without being aware of cosmic dynamics we're just beginning to understand. The beauty of the map is regardless of how we interpret the generic situation it represents, whether the interpretation is phenomenological, psychological, or ontological, boundaries still present themselves in that interpretation in a manner consistent with what's dictated by the first level of abstraction.

Given the fundamental nature of the first level of abstraction, what can it tell us about the progression of uncertainty at the second level of abstraction? If a quantification of uncertainty at the first level of abstraction is what we're calling a dimension, then a progression from subjective uncertainty to objective certainty at the second level of abstraction probably corresponds to a progression from higher dimensionality to lower dimensionality at the first level of abstraction. What we end up with is a dimension representing

chap 2 fn 5

a measure of uncertainty that's measured in dimensions. A dimension typically represents a measure—distance, time, pressure, temperature, etc.—that can be measured in some quantifiable way. What a dimension represents matters because dimensions that help us understand a situation, conceptual or real, are typically chosen to provide relevant information.

The easiest way to construct a conceptual framework upon the concept of dimensions is when each dimension is a fundamental pattern, such that the framework is a network of interacting fundamental patterns with the same morphology. If a pattern can be described as a relationship between concepts that reveals a potential functionality, then those concepts give us an idea of how the pattern can be applied to situations of interest. Unfortunately, a fundamental pattern's description doesn't tell us how it manifests itself in a specific situation, nor will it necessarily reveal specific situations a network of interacting patterns might present.

We're talking about two ways to use dimensionality: we might find it convenient to treat an appropriate quantification of uncertainty as a dimension, or a dimension can represent a pattern, a relationship between fundamental concepts, where potential functionality of the pattern is revealed by our choice of concepts. In the latter case a network of interacting patterns, a multidimensional construct, can represent a generic situation—a more sophisticated pattern—we can interpret as one or more specific situations.

5. Normally, the concepts of appearance and observation are associated with perception, but when they're not associated with perception—in the sense that an object can appear and be observed as something other than what appears and is observed as perception—the manifestations of appearance and observation may be unfamiliar. We can make the temporality of an object more intuitive by describing its appearance and observation as an expectation and a memory: the object as it appears, before it's perceived, appears as an expectation; and the object as it's observed, after it's perceived, is observed as a memory. The object observed as it appears in the now, the object perceived, isn't described as an expectation or a memory because it's an object that's neither an expectation nor a memory; it's an object in a temporal state, the now, where that state separates uncertainty from certainty.

6. Analysis of the interrelationship between the three uncertainties was motivated by Eddington's Fundamental Theory (Eddington 1949); however, I seem to recall a hard to obtain journal article Eddington authored, published prior to his posthumous publication, presenting an intuitive discussion of the uncertainties. The uncertainties of frame of reference, measurement, and observation might seem straight forward, but, given the unusual way these uncertainties manifest themselves in this presentation, their interrelationship might benefit from further reflection.

If you define the uncertainty of observation as the distance from the state of objective certainty to the state of subjective uncertainty, define the uncertainty of measurement as the distance from the state of subjective certainty to the state of subjective uncertainty, and keep uncertainty of the frame of reference defined as the distance from the state of objective certainty to the state of subjective certainty, then you have an intuitive relationship between the uncertainties where the uncertainty of observation is equal to the sum of the uncertainties of frame of reference and measurement without uncertainty of the frame of reference being a negative adjustment to the uncertainty of measurement. However, you now have uncertainties defined with respect to two different frames of reference with an uncertainty of observation dependent on a subjective measure of the object *as it appears*.

As we've presented it, the distance from the state of objective certainty to the state of subjective certainty is intuitive as uncertainty of the frame of reference because both states represent frames of reference. Is the distance

chap 2 fn 7

from the state of objective certainty to the state of objective uncertainty intuitive as the uncertainty of observation? Absolutely. The uncertainty of observation is uncertainty we want to be inherently objective; it should be measured from one objective state of the object as it's observed to another. The uncertainty of measurement supporting the uncertainty of observation, represented as the distance from the state of objective certainty to the state of subjective uncertainty, is consistent with the object-observed-as-it-appears presented as a metric spanning the same two states, but is it intuitive that the uncertainty of measurement should be adjusted by uncertainty of the range of that measurement? It is if the uncertainty of measurement already incorporates uncertainty of the frame of reference from which subjective measurement is made. This is what happens when the uncertainty of measurement is defined as measured from the state of objective certainty. Should the uncertainty of observation be a net or gross measure of uncertainty?

In both alternatives we've discussed in this footnote, the uncertainty of observation is defined from the perspective of objective certainty. Whether this uncertainty should be defined from objective certainty to objective uncertainty or all the way to subjective uncertainty is something we should think about. We can relate the alternatives to one another by proposing that you can assume uncertainty of what you observe from objective certainty equals the uncertainty of what you measure from subjective certainty (when one accounts for uncertainty of the frame of reference of the measurement) or you can assume uncertainty of what you observe from objective certainty equals the uncertainty of what you measure from objective certainty (when one accounts for uncertainty of the range of the measurement).

7. When our interest is the perception of an object by a perceiver, we traditionally favor an egocentric frame of reference—essentially a frame of reference located at the perceiver. For additional discussion regarding egocentric and

allocentric frames of reference, see (Meilinger and Vosgerau 2010; Klatzky 1998), and for a more detailed discussion of what classifies a frame of reference as intrinsic, relative, or absolute, see (Dokic and Pocherie 2006; Levinson 2004, chap. 2). You might also find the reference by (Musto 1999) of interest.

Although it seems natural to assume perception is egocentric, neither the perception centric perspective nor object centric perspective we're going to discuss in this section will be presented as egocentric. Later we'll suggest consciousness offers a sense of egocentricity even though it reveals itself as objectivity with respect to subjectivity as perception of (an object). The sense of egocentricity consciousness presents is inferred from the subjective frame of reference of its morphology. Regardless of whether we're examining the phenomenological dynamics of perception or consciousness, the intent is to maintain a separation between the phenomenological and psychological frameworks without abandoning the first person perspectives that make both frameworks of interest. Unfortunately, the first person perspective of either framework is less than obvious at the level of detail we're presenting.

Since we're accustomed to thinking about the perception of an object from the perspective of a perceiver's egocentric frame of reference, the idea of perception presenting itself as some kind of intermediary in the temporal progression of an object-observed-as-it-appears might seem less than intuitive. Perception is like consciousness in the sense that we can describe its functionality, the perception of something, or we can describe what it makes accessible, namely the object as it appears as perception and the object as it's observed as perception. Perception is always an intermediary.

8. Describing a measure of the object as it appears as an *expectation* and describing a measure of the object as it's observed as a *memory* are not intended as formal statements of the object's physical manifestation. The terminology offers convenient concepts chosen—in support of a theory—to

chap 2 fn 9-12

appeal to your sense of intuition with respect to the temporality of an object represented by the object-observed-as-it-appears.

9. You might want to compare what we're calling inner and outer horizons to Husserl's use of the same terminology (Welton 1999, 221–223). In our case, the inner and outer horizons are conceptualized to support and bracket our understanding of the now, in particular our understanding of the object-observed-as-it-appears as the object observed as it appears in the now encompassed by perception. For a more in depth discussion of Husserl's use of the terminology, see (McIntyre and Smith 1982, chap. V).

10. Close examination of figure 2-3 reveals a context where the states of subjective certainty and objective uncertainty in the now can be claimed to have the same measure of uncertainty with respect to the *certainty* frame of reference. If you ignore all states associated with perception, what remains is a representation of the object-observed-as-it-appears by four states such that the above states, as the object observed as it appears, defines the now.

11. The conceptual framework upon which our phenomenological interpretation is based favors a direction, from the subjective to the objective, expressed by a situation described as consciousness of (perception of (an object)). The inner horizon, a representation of perception of (an object), is encompassed by consciousness where objectivity is realized when the perception of an object, an object observed as it appears in the now, collapses to its expression as consciousness. The implication is the object as it appears as an expectation and the object as it's observed as a memory are outside of the field of perception. They can only be brought into the field of perception as a manifestation of an object in the temporal now.

12. We've presented perception as an intermediary that enables a temporal progression, from uncertainty toward certainty, between what will turn out to be two transcendent states of an object. We used figure 2-3 to demonstrate that perception has an objective and subjective relationship with an object that combine to present the appearance and observation of that object as perception. What ensures that the object as it's observed as perception is the same object as it appears as perception is consistency with the object observed as it appears in the now. The object-observed-as-it-appears, however, reminds us that the object observed as it appears in the now must be consistent with the object as it appears as an expectation and the object as it's observed as a memory. This is by design. The object as it appears as an expectation and the object as it's observed as a memory are measures chosen to support the subjective apprehension of an object. (See footnote 8.)

13. While a progression of the object-observed-as-it-appears, from subjective uncertainty to objective certainty, is convenient to our understanding of the perception of an object, we have to be careful about assuming the goal of perception of (an object) is realization of the object-observed-as-it-appears as objective certainty described by the object as it's observed as memory. Perception of (an object) conceptualizes the subjective side of a situation where the object-observed-as-it-appears represents a metric required to make the conceptualization intuitive. The goal of perception of (an object) is likely achieved when the object observed as perception is the object as it appears as perception, suggesting the object-observed-as-it-appears is the object observed as it appears in the now.

In chapter three, patterns of states will be interpreted to help us understand the phenomenological situation described as consciousness of (perception of (an object)). The potential we've been describing as subjective perception of (an object) will be actualized as objective consciousness where consciousness is represented by a progression of states, from a state of subjective uncertainty of consciousness to a state of objective certainty of consciousness. Although it will be tempting to assume the goal of consciousness of (perception of (an

chap 2 fn 14-15

object)) is a measure representing the objective certainty of consciousness, we might be better advised to start with the proposition that the goal of consciousness of (perception of (an object)) is consciousness. We can subsequently reflect on why consciousness might be compelled to collapse to a measure representing objective certainty and whether such a collapse is possible.

14. One could argue that the object as it appears as expectation is expectation of the object-observed-as-it-appears as the object observed as it appears in the now, and the object as it's observed as memory is memory of the object-observed-as-it-appears as the object observed as it appears in the now. In this sense, the object observed as it appears in the now, what was expected and what's to be remembered of the object-observed-as-it-appears, could be claimed to be the essence of the object-observed-as-it-appears.

The now as we've defined it only accommodates the object-observed-as-it-appears as the transcendental object observed as it appears. We've outlined a description of the object-observed-as-it-appears where measures of the metric defining an outer horizon, encompassing the inner horizon and the now, are recognized as transcendent. Ideally, we'd like the now to accommodate the entire range of the object-observed-as-it-appears described as transcendental and transcendent. To achieve this accommodation, we either expand our definition of the now to encompass the outer horizon of the object-observed-as-it-appears or find a way to collapse the outer horizon such that it's encompassed by the now as we've defined it in the inner horizon. Which option is taken and how that option is achieved is addressed in the primary document.

15. The transcendent measures of the object-observed-as-it-appears are proposed as transcendent to perception of (an object) represented by the inner horizon in figure 2-3. We've described these transcendent measures as an expectation and a memory because they represent temporally intuitive descriptors intended to communicate the progression of states of the object-observed-as-it-appears from uncertainty to certainty. The object-observed-as-it-appears as the object observed as memory is not necessarily memory of experience of an object, and the object-observed-as-it-appears as the object as it appears as expectation is not necessarily expectation of experience of an object. The object-observed-as-it-appears as the object as it appears as expectation describes an object before the now as subjective uncertainty with respect to its appearing as subjective certainty in the now, and the object-observed-as-it-appears as the object as it's observed as memory describes an object after the now as objective certainty with respect to its being observed as objective uncertainty in the now. What appears to be missing is the connection between the object as it appears as expectation and what we anticipate to experience with respect to that object. Needless to say, the same claim can be made with respect to the connection between the object as it's observed as memory and what we retain of experience with respect to that object.

Chapter 3

1. Claiming we're going to engage in phenomenological analysis in this presentation would be a bit of a stretch; that task is undertaken in the primary document. What we will do is present a situation, namely consciousness of (perception of (an object)), inviting understanding in the context of phenomenology without engaging in details of phenomenology as a methodology or philosophy. The epistemology of the situation is also addressed in the primary document.

2. Since the discussion is presented with a focus on the second level of abstraction, it's important to keep in mind that the first level of abstraction is fundamental to the geometry upon which the second level is formulated.

3. See (Goodfellow 1987; Legon 2005).

chap 3 fn 4-5

4. Also see Jim Alison's "Proposed Design Specifications for the Great Pyramid and for the Siteplan of the Giza Plateau."

5. In order to progressively build upon Goodfellow's original insight regarding the vanishing point, we need to start with a reasonably close representation of Legon's illustration. (There's no intent to suggest our computer generated representation is more accurate than the illustration produced by Legon.) To keep the figures comparable, our diagrams were tweaked to best accommodate relative positions of the pyramids with respect to each other, certain non-observables, and the vanishing point.

There may be a small discrepancy between Legon's figure (see appendix 2) and our representation of the vanishing point as presented in figures 3-1 through 3-3. In figures 3-1 through 3-3, point O is the center of the circle encompassing the pyramids such that the line segments ON and OL are not exactly 90 degrees with respect to each other; the angle is closer to 87 degrees. In Legon's figure, those line segments appear to be very close to but perhaps not exactly 90 degrees apart. For the purpose at hand, differences between the representations don't appear to be significant, but we're inclined to make the assumption that any discrepancy from Legon's figure, presumably representing a deviation from what's observed, is a discrepancy we've introduced. When we introduce figures 3-6 and subsequent figures, you'll note they assume line segments ON and OL are ninety degrees apart.

(The central figure in chapter three is figure 3-6 because it's the figure most representative of the conceptual framework mapped by the pyramids. All of the observations presented in figures 3-1 through 3-14 can be recognized by reflecting on figure 3-6. For presentation purposes, however, we've chosen to start with the vanishing point observation illustrated by figure 3-1; it represents a familiar stepping stone offering the historical context of a documented observation that seemed to have no obvious explanation. It would have been disappointing if the vanishing point observation suggested by Goodfellow and verified by Legon had not turned out to be significant.)

6.   The reason point K has been added is to suggest it's a point the pyramids are intended to bring to our attention. Point K is one of the more obvious points we've discovered warranting that kind of attention.

7.   The desire to introduce point K is the primary motivation for presenting our own illustration of Legon's original figure. Understanding the importance of point K beyond its relationship to the pyramids takes us further down the rabbit hole than we're inclined to address, but point K is of immediate interest as a possible explanation for the angle of inclination of the Ascending Passage of the Great Pyramid. That angle accounts for the compromise in the location of the King's Chamber. (See (Legon 2000A) for a calculation of the angle of inclination and comparison to measures of that same angle reported independently by Petrie and Smyth. All of the documented measurements are close to the value of the angle (25.99 degrees) in figure 3-3.) The Ascending Passage is one of the most intriguing features of the Great Pyramid.

8.   If you're having difficulty reconciling figure 3-1 with figure 2-4, it's important to understand that figure 2-4 illustrates subjectivity as perception of (an object). Figure 2-4 can only accommodate the objectivity of consciousness as a conceptualization. Figure 3-1, on the other hand, illustrates the relationship between subjectivity as perception of (an object) and objectivity as consciousness with the intent of communicating the idea that perception of (an object) when that object is observed as it appears is somehow transformed into consciousness: it attempts to associate objectivity as consciousness, represented by a circle centered along the cone of actualization, with subjectivity as perception of (an object), represented by three pyramids. Later in this chapter, a different figure–figure 3-6 to be exact–will present consciousness as points along the axis of the cone of actualization. In that

chap 3 fn 9-11

figure, we'll be inclined to represent subjectivity as perception of (an object) in the form of a conceptualization also presenting itself as points along the same axis.

9. Actually, the square at point D extends slightly beyond the boundaries of the cone of actualization. Although there's a version of this figure where the square at point D lies within the boundaries of the cone of actualization, the situation underlying the figure shown has unique relationships in its geometry we find appealing.

10. Illustration of the relative sizes and locations of the three pyramids within the squares drawn at points C and D in figure 3-4 and subsequent figures represent visual approximations that are sufficient for the level of accuracy needed in this presentation.

11. In our recreation of Goodfellow's original insight, presented in figures 3-1 through 3-3 in a manner consistent with Legon's figure, the axis of the cone of actualization (at 44.89 degrees) is not exactly 45 degrees, the line segments from point O to points N and L are slightly less than 90 degrees with respect to each other, and the lines drawn from point K to the center of the smallest and largest pyramids create an angle of 25.99 degrees. Figures 3-4 and higher by contrast are based on an axis of the cone of actualization at exactly 45 degrees, line segments ON and OL (see figure 3-8) at exactly 90 degrees apart with respect to each other, and a significant difference in scale (see figure 3-6) when compared to figures 3-1 through 3-3.

In figure 3-6, we did find it necessary to readjust relative positions of the pyramids with respect to each other while trying to keep the line segments from point K to the center of the smallest and largest pyramids close to an angle of 26 degrees with respect to each other. That angle in figures 3-6 through 3-8 is 26.05 degrees. (Elaboration regarding readjustments might be worth documenting for future revisions: Figures 3-4 and higher were all originally created using the same base diagram, which means a discrepancy in the base diagram would be reflected in all figures created from it. The decision to adjust the position of the middle pyramid at point B to target a 26 degree angle was made as a post-production edit after discovering the relative position of the middle pyramid in the base diagram was drawn slightly out of alignment with respect to the other two pyramids. Figures 3-4 through 3-6 were subsequently recreated from the same adjusted base diagram. The remaining figures requiring adjustment were adjusted separately with imperceptible discrepancies between figures. Fortunately our interpretation for most of this chapter isn't dependent upon the exact positioning of the pyramids with respect to each other. The positions of the pyramids at points C and D, for example, are based on visual approximations.) Finally, the choice of rendering the axis of the cone of actualization as a 45 degree line was a matter of convenience.

12. What figure 3-8 demonstrates is some assumption relating figure 3-1 to figure 3-6 allows us to recognize figure 3-1 as embedded in figure 3-6 such that point O and point B are nearly coincident. The downside of the assumed relationship is we can't adjust figure 3-6 such that points O and B coincide simply because it's convenient to the case we're trying to argue. So either the assumption–whatever it is–isn't exactly correct, or there's a reason the vanishing point relationship in figure 3-1 wasn't designed to align where we would have liked with respect to figure 3-6.

In our interpretation, we're going to assume the circle near point B is centered exactly at point B, i.e. points O and B are coincident. We're making an assumption that the circle not being centered exactly as expected at point B doesn't alter the general interpretation. (See comments at the end of appendix 3 for additional speculation regarding what the discrepancy might reveal.) There's no need for concern regarding this assumption since

chap 3 fn 13-14

our interpretation is formulated from frameworks presented in the primary document; however, what we're attempting in this presentation is a reconciliation of the map presented by the pyramids at Giza with that interpretation. In other words, whether points O and B are coincident or not doesn't alter the general interpretation; what it does affect is the accuracy by which the map presented by the pyramids can be claimed to represent our interpretation.

13. We're confident in the approach of starting with perception of (an object) and ending with consciousness, but details of how the transformation unfolds at this level of interpretation are less clear. Is the transformation from perception to consciousness progressive or instantaneous? Claiming perception of (an object) collapses to consciousness is intuitive as an instantaneous collapse at all four points along the cone of actualization because the collapse to consciousness, as we've described it, requires an event described as the perception of an object observed as it appears in the now. It's also conceivable that the collapse and the event occur simultaneously. Alternative possibilities regarding the timing of the collapse may come to mind when you encounter figures 3-9 and 3-10. Those figures are intuitive with respect to visualizing the collapse to certainty of consciousness of (perception of (an object)), but they have their limitations. Regardless of how either collapse occurs, we can't dismiss the likelihood of an intimate connection between measures of consciousness and perception revealing itself as a network of intentionality that enables the phenomenological apprehension of an object.

14. What's interesting about figure 3-6 is we seem to have a pattern within a pattern: the three circles and the solid squares representing consciousness of (perception of (an object)) at points B, C, and D are analogous to the three pyramids at Giza, and consciousness of (perception of (an object)) realized at the certainty frame of reference at point A is analogous to the vanishing point. The analogy implies consciousness of (perception of (an object)) located at point C represents the middle pyramid, suggesting it might also represent the now in this pattern. This is the same location where the coincidence of solid and dashed squares suggests perception of (an object) accommodated by the area of the solid square represents the maximum of what can be accommodated by the area of the dashed square. The area of the dashed square represents the amount of perception of (an object) it's possible to accommodate by something we've yet to identify. The observation is certainly compelling, but just how much we should read into this analogy, or try to extract from it, is less than obvious. Thus far, we've associated the layout of the pyramids and their vanishing point with subjectivity expressed as the perception of an object. Points A, B, C, and D, however, are associated with objective consciousness that obtains from the collapse of perception of (an object). It may be a minor point but the analogy vanishes with the collapse of subjective perception of (an object) to objective consciousness. Because of its visual appeal, this is an analogy we want to keep in mind.

15. Figures 3-9 and 3-10 give us the opportunity to note that even though we've proposed an additive relationship between the uncertainties of observation, frame of reference, and measurement underlying consciousness represented by points A, B, C, and D along the axis of the cone of actualization–uncertainties represented as *distances* measured with respect to a certainty frame of reference at point A–we've made no such claim of an additive relationship between the subjectivities of perception of (an object) represented by *areas* of the solid squares encompassing three pyramids at each of points A, B, C, and D. In other words, assuming points A, B, C, and D represent measures of objective consciousness positioned with respect to the certainty frame of reference at point A, even though the measure of uncertainty representing the uncertainty of observation of consciousness (at point C) is equal to the measure of uncertainty representing the uncertainty of

chap 3 fn 16-18

measurement of consciousness (at point D) negatively adjusted by an amount equivalent to the measure of uncertainty representing the uncertainty of the frame of reference of consciousness (at point B) the same additive relationship isn't claimed for subjectivities of perception of (an object) represented by the solid squares at the same points.

16. The scenarios in figures 3-9 and 3-10 are intuitive, but you can't assume they're definitive. Because they're visual, these figures might inspire insight into other ways of representing figure 3-6 capable of accommodating a collapse to certainty.

17. You may be wondering why we took the time to introduce a scenario knowing it couldn't be a viable option. The entire presentation is a guided tour designed to eliminate all of the countless hours of groping in the dark one has no choice but investigate. The intent of introducing this scenario is communicating that many of those options were investigated. It never hurts to speculate as to what you might be missing.

18. In figure 3-6, consciousness encompasses perception of (an object) and perception encompasses the object observed as it appears in the now. We've also suggested a sense of temporality with respect to the now we can describe as the object as it appears as perception before the now and the object as it's observed as perception after the now. Should we expect an analogous temporal sense of consciousness with respect to the now we can describe as perception of (an object) as it appears as consciousness before the now and perception of (an object) as it's observed as consciousness after the now? The temporal sense of consciousness we've described would be a conceptualization with the virtue of consistency, but that conceptualization is short-circuited by a situation where the collapse of subjective perception of (an object) to objective consciousness collapses the subjective now. Details of how consciousness presents a sense of temporality are less than obvious.

chap 3 cont fn 18

There's a sense of synchronicity between the temporal *progression* of perception of (an object) and the *transformation* from subjective perception of (an object) to objective consciousness: the progression is a prerequisite to the transformation, arguably a structural requirement, implying some sense of synchronization should be expected with respect to the now. In the case of perception of (an object), synchronization with the now is illustrated in figure 2-3. In the case of consciousness of (perception of (an object)), illustrating the synchronization with the now, see figure 2-4, isn't as straight forward. We should, however, not be surprised that the transformation from perception of (an object) to consciousness requires a now defined by the object observed as it appears, essentially an event made possible by perception.

Although it's conceivable that perception can be transformed into consciousness, transforming the now along with it is harder to visualize, particularly since it's possible that the object observed as it appears defining the now represents the most likely means of connecting the phenomenal with the noumenal. If perception of (an object) collapses to consciousness, what happens to the object observed as it appears defining the now? What ontological status allows the now to collapse in the first place? Fortunately, the very concept of the now—commonly understood as an abstraction—is inherently ephemeral, which means the idea that the now is something that can collapse doesn't require a great leap of faith. Transforming subjective perception of (an object) into objective consciousness probably facilitates representing the same information in a more utilitarian form that's easier to articulate or comprehend.

When perception of (an object) collapses to objective consciousness, is there anything inherent in consciousness suggesting the experience it makes accessible is identified as the experienced present? Is consciousness of an object, particularly consciousness of (perception of (an object)), only identifiable as consciousness of an object observed as it appears in the now because of how we've defined the situation? What makes the now immediate, and how does the immediate now differ—if it does—from the now just passed and the now soon to be? These aren't trivial questions, and even more questions surface when we move beyond the now and attempt to understand the consciousness of time (Welton section 11). What is the relationship between consciousness and time? If time isn't a progression of nows, what is it? What synchronizes consciousness of (time) with consciousness of (perception of (an object))?

There's another sense of temporality we should at least mention. At some level, the dynamics of *temporality as a way of being* might be analogous to the dynamics of our phenomenological situation. To fully appreciate how the conceptual geometry, a geometry upon which the generic situation and its interpretation as a phenomenological situation is built, might accommodate temporality as a way of being—a *running ahead* to the certainty of *authentic historicity* of an entity—requires reflecting on what Heidegger describes as the *authentic temporality* of *Dasien* (Massey 2015, chap. 1, sec. 4; Gorner 2007, chap. 8; Guignon 1983, chap. III).

Heidegger uses Dasien as a technical term to refer to human life or existence. Dasien must progress temporally from *situatedness* in a world, what Heidegger describes as essence, to *projection* towards its ability-to-be as existence. Assimilating Heidegger's contribution into a situation supported by the map might be easier than it seems if you focus on the temporality of Dasien and reflect on what that temporality is expected to achieve. A complete understanding of the authentic temporality of Dasien probably requires its own framework, but it can be done. (Is Dasien analogous to the object-observed-as-it-appears, or is it analogous to perception? If we assume the former, then an obvious first cut might be the analogy of authenticity of (Dasien) with perception of (an object). If we assume the latter, then the analogy of Dasien of (*something*) with perception of (an object) hinges on what *something* represents.)

19. Try to keep in mind that because the solid squares are located at points along the axis of the cone of actualization corresponding with uncertainties of observation of underlying sub-patterns of the situation illustrated in figure 3-6, the size of each solid square increases in proportion to the uncertainty of observation at each point.

20. The distance from point B to point G is equivalent to the distance from point A to point F representing uncertainty of the frame of reference of this sub-pattern. The equivalence is necessary because the distance from point B to point G represents a negative adjustment to the uncertainty of measurement (see footnote 6 in chapter two) allowing uncertainty of the frame of reference plus the uncertainty of measurement to equate to the uncertainty of observation defined by the sub-pattern.

21. Think of each sub-pattern as a set of measures of uncertainty of frame of reference, measurement, or observation. Measures representing the distance, from point A, described as the uncertainty of observation in each sub-pattern (measures of objective uncertainty) are points A, B, C, and D. These are the same points that define the sub-pattern representing the uncertainty of observation of the super-pattern. A similar claim can be made that points A, F, B, and G representing the distance, from point A, described as uncertainty of the frame of reference in each sub-pattern (measures of subjective certainty) are the same points defining the sub-pattern representing uncertainty of the frame of reference of the super-pattern. A claim can also be made that points A, G, D, and J representing the distance, from point A, described as the uncertainty of measurement in each sub-pattern (measures of subjective uncertainty) are the same points

chap 3 fn 22-23

defining the sub-pattern representing the uncertainty of measurement of the super-pattern.

22. In figure 3-6, we can interpret each dashed square as a representation of subjective perception of (an object) that can be accommodated at the uncertainty of observation associated with consciousness, which is why the dashed and solid squares are coincident at point C. (Recall that the distance from point A to point C represents the uncertainty of observation with respect to consciousness represented by points A, B, C, and D.) What appears obvious is this limit to accommodation happens to be why the three pyramids at each of points A, B, and C in figure 3-6 occur within the boundary of the dashed square at each point. The three pyramids at point D, by contrast, occur outside of the boundary of the dashed square at that point. At point D, subjective perception of (an object) represented by the solid square is greater in area than what can be accommodated at the uncertainty of observation of consciousness represented by the dashed square at point C, conveniently repeated at point D. How much subjective perception of (an object) the phenomenological situation is capable of accommodating appears to be constrained by how much of it can be resolved at the uncertainty of observation of consciousness. Is there a relationship at the first level of abstraction making the constraint intuitive?

23. The distinction between inner and outer horizons appears necessary when we limit our perspective to subjectivity as perception of (an object). The need for that distinction isn't as obvious from the perspective of the entire situation described as consciousness of (perception of (an object)), yet it seems you can't have the latter without accommodating the former. The former is apprehension of the object-observed-as-it-appears in its progression from uncertainty to certainty, and the latter is a transformation where the object-observed-as-it-appears must be perceived as the object observed as it appears in the now for it to be capable of collapsing. The problem with contemplating elimination of the distinction between inner and outer horizons is you probably can't eliminate the distinction between transcendental and transcendent objects–what essentially defines the horizons. The easier alternative might be to find a way to eliminate the need to accommodate a temporal progression of the object from uncertainty to certainty.

You can't try to comprehend the phenomenological situation without taking quantum physics into account. The situation suggests the need for a change in the ontological status of an object in a manner the temporal metric described as the object-observed-as-it-appears can't accommodate if the object observed as it appears in the now is only transcendental. According to quantum physics, consciousness of (perception of (an object)) when that object is observed as it appears in the now should necessitate a transcendent object found in the now prior to collapse of perception of (an object) to consciousness. The question to be asked is why doesn't it? If a transcendent object *is* found in the now as a result of its being observed as it appears in the now, what ontological status of that object might allow it to collapse as part of the collapse of perception of (an object) to consciousness?

24. Before we move on, we should note that the angle of the face of the Great Pyramid is an exacting requirement of the conceptual geometry best appreciated by analysis at the first level of abstraction. Although we won't examine it here, it's an analysis suggesting the angle has an importance justifying its illustration by the Great Pyramid and is presumably why the Great Pyramid was designed to bring it to our attention.

25. As an approximation to this location, note that $\tan(20.442°) = 82$ cubits/220 cubits = 0.3727. See (Legon 2000A) for cubit measures.

26. It is not surprising that the King's and Queen's Chambers and their respective air shafts might be subject to multiple interpretations with respect

chap 3 fn 27-30

to the conceptual frameworks. For the presentation at hand, we're going to limit our attention to one interpretation.

27. This is typically an engineering debate where the primary focus is what's required to ensure integrity of the structure.

28. The five levels of the niche can be argued as representing a *focusing* of five levels of understanding of the situation described as consciousness of (perception of (an object)). We aren't presenting the details here because the insight is speculative, and we need the frameworks of the primary document to adequately explain the dynamics of how the focusing might be achieved.

29. Even though the golden pyramid represents an approximation to the Great Pyramid, we'll continue to refer to point B in figure 3-11 and to related points in figure 3-14 as GPs. It should make reference to points of interest easier.

30. Let's revisit a question addressed at the beginning of this presentation. Why would extraterrestrial visitors care enough about the future of human civilization to bother leaving us a message? If extraterrestrials cared enough to leave us a message, then it's possible that they're responsible for our being on this planet in the first place. Another hopefully remote possibility is life in the universe is so rare that the discovery of any form of life, primitive or advanced, is an occasion to be celebrated. The discovery of a planet capable of sustaining life and subsequently finding sentient life on that planet—our planet—compelled those visitors to leave us a clue saying, "You're not alone." Implicit in the clue is the idea that whoever left it figured they might not be here to tell the story; they either moved on or passed on. Yet another possibility is the message represents a testament to a search for identity and meaning from visitors clearly highly advanced but also a lot more like us than we imagine. The desire to pass on knowledge, tell stories, and understand one's place in the universe may not be unique to humanity. How did they know that the message they encoded would represent knowledge we struggle to understand? Were they attempting to pass the baton of wonder, or were they communicating an apprehension? Was the message nothing more than an expression of vanity? What exactly is the message?

Finally, imagine that some members of a team of explorers might have chosen not to return home, all for the sake of more freedom. Imagine the extraterrestrials might have been refugees who, out of necessity, wandered here in search of a habitable planet. These are unlikely scenarios but certainly not outside the realm of possibility.

31. Although you might think the primary document picks up where this presentation ends, it's not the case. The two presentations are connected conceptually by their mutual dependence on the same underlying conceptual geometry; however, the primary document was written entirely at the third level of abstraction as a self-contained development of the frameworks. It was written before this presentation. The primary document develops the conceptual frameworks to incorporate psychology, phenomenology, and ontology. It also develops the phenomenological situation in more detail. The primary document doesn't concern itself with extraterrestrial contact or the pyramids at Giza, but it is concerned with what's fundamental in the order of things particularly as it manifests itself in ways we assume are uniquely human. Evidence presented in one document lends credibility to the other document.

32. What comes to mind when we speculate regarding organisms of our own creation is artificial intelligence and marrying artificial intelligence with robotics. While engineering a machine to match the intelligence and dexterity of a human organism gives every indication of being a matter of engineering, the possibility of that machine exhibiting consciousness and a sense of self isn't as obvious. The biggest obstacle we face in developing an

chap 4 fn 1-3

intelligent machine capable of consciousness and a sense of self, on par with a human organism, is our basic lack of knowledge of how a human organism achieves these capabilities and of how the two are married together. We're talking about two of the great mysteries of humanity. What's the connection of a sense of self to the organism as subject of its consciousness? In order for an organism to claim consciousness of its consciousness of an object, must an organism have consciousness of itself? Does self-consciousness necessarily lead to an infinite regress? See (Neuhouser 1990, chap. 3, particularly pages 73 and 74) for a statement of the problem.

We haven't discussed the self or how it relates to consciousness because the discussion requires more than the phenomenological situation can accommodate. The beauty of it is additional frameworks map to the same conceptual geometry we've been discussing.

Chapter 4

1. One of the things you give up when you step outside of a career development environment is the benefit of mentoring. Mentoring is important in a corporate environment because the technical challenges often pale in comparison to the political challenges. Developing MIS as an employee of a corporation often puts you in a precarious liaison position.

2. An MIS project that takes more than three years to develop typically ends in failure, or it's obsolete before it becomes operational. (There are many reasons why MIS projects fail: the software isn't flexible enough, the data isn't provided accurately or timely enough to be of value, too many marginally beneficial reports, and everyone seems to have their own interpretation of what the facts are telling them, just to name a few.)

3. I don't recall criticisms referring to my apprehensions as crazy, but I do recall one criticism referring to them as naïve. That criticism was leveled by someone in a tenured teaching position for whom labor was just a variable in an equation. There was truth to that criticism. He was referring to the sense of inevitability that comes from realizing you're facing a rising tide of technological innovation and globalization giving every indication of being unstoppable. If anything was naïve it was assuming people really wanted to hear the truth. The problem with hearing the truth is you have to face the truth, and the problem with facing the truth is you're compelled to do something about it. Most people seem content to convince themselves they'll be survivors; they ride the train for as long and as far as possible.

4. Were corporations really so out of touch that armies of analysts were incapable of predicting the financial impact of what was unfolding? Projections presented to corporate boards don't necessarily reflect results of the budgeting process or economic reality. Executive compensation and longevity can be linked to salesmanship or occasional accounting tricks. What can happen in some situations is a division president gambles on forecasted sales and profitability being wrong or memories being short. In the meantime the board is hopelessly uninformed about reality.

5. Thirty years later you might think this is all ancient history, but we're talking about a long-term economic trend yet to run its course.

6. Speaking of pedestrian, I remember how one professor—in the middle of an undergraduate lecture—decided to offer a lesson in life when with a hint of lamentation behind his smugness he calmly advised, "Never strive for the pedestrian in life." I don't know why he said it, but it stuck with me even though I can't remember what the lecture was about, the class being taught, or the name of the professor. I also remember sharing beers with some contemporaries in graduate school where I mentioned a cosmological insight revealing itself while studying statistics. That recollection prompted one contemporary to shake his head saying, "You're certainly not afraid to

chap 4 fn 7-9

ask the big questions." And I wasn't. Unfortunately, those insights had a nasty habit of popping into my head when I really needed to be focusing on something else. The problem with trying to expand your horizons is sometimes the things you see over those horizons can haunt you. As it turns out, the insight recollected in graduate school would end up playing a key role in my later research.

7. I felt compelled to fall back to the more comprehensive perspective of general systems theory so I started by pulling out my copy of (Bertalanffy 1969).

8. With respect to your environment, you generally have three choices: adapt to the environment you find yourself in, find an environment where you can thrive, or create your own environment. On the surface of it this statement appears fairly innocuous, yet it can have profound personal and social implications. In this country, for example, our biases toward certain members of society coordinate to keep them trapped in an environment of limited opportunities, yet when those members of society adapt to their environment in ways that shock us we blame them for not doing more to escape the very environment we helped to create. The mass migration of young people away from the countryside and into urban centers represents an attempt on their part to find an environment where they can thrive economically. The lure of social media is the lure of an environment in which participants feel the need to belong. Finally, you have the case of extremists deciding to create their own environment; this is where consequences can get dicey if creating one's own environment is accomplished at the expense of someone else's environment.

9. I initially called it Resonating. In retrospect, Synthesizing is probably a more appropriate description because what I was essentially doing was trying to synthesize information in every direction at once in a deliberate attempt at incorporating what seemed fundamental from the perspective of different disciplines. Synthesizing required some imaginative mental creations.

10. The correspondence described predated the giant vanity press we know as the Internet. Nowadays, members of the academic community may receive marketing material intended to promote the commercial potential of the frameworks, but we don't send individuals or organizations unsolicited manuscripts.

11. For better or worse, our sense of self isn't achieved in a void. We develop our sense of self in the context of a world with participants that project an essence upon us. We can try to rise above it all but the milieu is always there. If we invest too much of ourselves in cultural, religious, or economic contracts that begin to break down, it should come as little surprise that some form of reconciliation becomes necessary.

12. People can't feel like they matter when their basic economic needs aren't being met, but it takes more than simply helping them survive: they need to feel like they're being treated fairly, and they need assurance the system isn't rigged against them. People need to feel like they're part of something offering some sense of meaning to life, to feel like they're making a contribution to something bigger than themselves, and to feel the world will be a better place as a result of their being a part of it. People need inspiration and hope. We all have these needs in common; when they're not being fulfilled we can't live up to our potential as human beings.

13. This reinvention wasn't the first time I felt it necessary to take dramatic action to alter the course of my life. In the mid-seventies I decided to hit the reset button and bring conflicting priorities, undisciplined behavior, disenchantments, and haunting loneliness to a dramatic halt. I needed to regroup and attempt what's commonly described as some soul searching. It's not a

chap 4 cont fn 13

coincidence that the decision was made after returning from a two month trip to Egypt and its antiquities. It was an abrupt decision that would prove costly in many ways but it was necessary. There was something missing in my life and resetting everything was the first time I made a serious effort to figure it out. One of the benefits of that reset was the decision to change course and pursue a career in computer science—a timely decision because I desperately needed to pursue a goal of genuine interest. Who would have guessed computers would become ubiquitous?

(Experiencing more of the world and its different cultures can alter your world-view and where you see yourself in unfolding revelations about the world. More often than not, it's a humbling experience, particularly when you're young and naïve and forced to make choices in a world in which you're compelled to adapt. In a modern industrial society, you're compelled to keep swimming upstream and ignore the certainty of death, yet faced with that certainty you're inclined to question how you should live. You begin to question what's you versus some conditioned response fitting a narrative you're wearing like the latest fashion. You keep grasping for what's real only to find that everything real lives in a fleeting now.

Although there's safety in numbers, tribalism comes at a price requiring each of us to conform in behavior or world-view—sometimes to the point where it becomes stifling to our development as individuals. You can find yourself living in a social reality among influencers with radically different world-views—worlds within a world—all projecting their values upon you. Sooner or later you begin to question whether the world you're living in is a thin veneer layered over a reality no one seems inclined to embrace, a world in which you question the price it demands for the refuge it provides. What happens when your world-view becomes dangerously divorced from reality? What can happen is reconciliations with reality require dramatic changes to be effective. Do you embrace those reconciliations or continue to entrench yourself in the shrinking world of your own delusions?)

So, if I may change the subject, what are we to make of a map that gives every indication of being a description of the human situation as a quest for certainty in a Cartesian world of a solipsistic subject, defined by a mind, surrounded by external objects? Does my consciousness of (perception of (an object)) depend in any way on the phenomenological situation realized by another such as me? What can the map tell us about how interaction with other subjects helps to define us or helps us understand the world we see around us? Is the phenomenological situation sufficient to provide additional insight into our interactions with others, or do we need to call upon other situations built on the same conceptual geometry? Must we project something of ourselves upon others like ourselves–creation of an alter ego–to understand how we relate to each other, or is some form of mutual interaction at play that's more fundamental than we realize? I'm talking about an idea in phenomenology called *intersubjectivity* (Zahavi 2003, chap. 3; Welton 1999, chap. VI; Lauer 1978, chap. 8; Moran 2000, chap. 5; Frie 1997). Does the map present any clue to the nature of intersubjectivity? Why did Husserl consider intersubjectivity so important to transcendental phenomenology?

14. The important contributions of the GOP research to my later research were reflections regarding the centrality of time and assumptions about the nature of time reflected in investment decisions. (Early in that project I wrote a working paper titled "The Time Relativity of the Discount Factor," and I used that working paper to develop a model for valuing securities.)

15. My most enjoyable attempt at making a living since the early eighties–something more than just a job–was a brief detour into the arts where I was technically a partner in a partnership. The idea was to try a project where I had a chance to pay more attention to what I was feeling. Even though I lost

chap 4 fn 16

my shirt on the venture, it demonstrated that collaboration could be a very rewarding experience, and it taught me the value of creating products representing finely crafted, well distilled, gems—not so easy when you're working on a budget in your spare time. Producing creative content, particularly in a collaborative environment, was often very consuming, and it required a great deal of diplomacy and mutual respect to make it an enjoyable experience. Interestingly enough, the arts involve recognizing and interpreting patterns in novel ways that inspire and appeal to our senses, emotions, sense of self, and sense of belonging.

16. Corporate environments were changing so dramatically and so quickly that my approach to employment would ultimately impact my approach to research. I was developing a sense of paranoia regarding prospects for long-term employment as I observed the progressive commoditization of labor and dramatic job losses. Those observations impressed upon me the need to learn how to do as many different things as I could within the corporate environment, even if I had to learn at the corporation's expense. It turned out to be a humbling experience, in more ways than one, but in each experience I was also developing varied and informed insight regarding the nature of organizations in the context of a capitalistic sociopolitical economy. Along the way, I was also discovering revealing insights about myself.

I learned that taking a job just for the sake of having a job was always going to be a losing proposition for me and my employer. I simply wasn't the kind of guy who could suck it up in a job when I really didn't want to be there. I tried my best to adapt to those environments, but, most of the time, I didn't like what I was becoming. Eventually I caught on to the fact that if corporate work didn't involve the development or application of new technology, I had a hard time getting excited about it. Even the work you're currently reading is motivated by the pursuit of artificial consciousness.

With respect to organizations, each experience impressed upon me that to truly understand an organization, organism, or environment you have to understand it from a number of perspectives. Models employed to help distill what you observe, whether they're implicit or explicit, necessarily leave something out. Things you tend to ignore are things you least understand or appreciate. You really have to be careful not to get yourself stuck in some conceptual silo convenient to your ignorance regarding the complexity of what you're confronting. What tends to happen is you dismiss what you don't understand or don't value as being unimportant. The problem is what's unimportant under one set of conditions can become critical as conditions change. What you're observing is typically telling you everything you need to know if you've invested in what's required to fully comprehend it and you observe it for a sufficient period of time.

17. I really enjoyed the wonderful libraries made available by major universities. My favorite pastime in college and many years after college was escaping from the world into the stacks of those libraries to peruse specialized journals and follow reference trails in pursuit of articles with those rare qualities capable of making my respite all the more rewarding. Perhaps it was my own confusion but I often felt some fields of study could benefit from a little more cross disciplined cooperation and collaboration. You can learn a lot about the strengths and weaknesses of any theory or paradigm by attempting to reconcile it with a body of knowledge in another discipline with different priorities. Even within a particular discipline, the true test of any new theory is its ability to account for accepted theory. If you read enough journals, in enough fields of study, and attempt to assimilate all of that information, you can't help but look for common patterns capable of putting everything into perspective.

(Consider how many times the word *pattern* has been invoked in

chap 4 fn 18-19

this presentation in an attempt to communicate a common thread in our understanding. The world we inhabit is one of patterns built upon patterns, a world where what patterns have in common is more important than how they differ. Recognizing patterns is important, but it won't do you any good if you don't understand what they're trying to tell you. The things patterns can tell you depend on the context in which they manifest themselves, contexts that are more interconnected than they appear.)

18. The game was a Chinese board game known as GO. GO was a popular tool for AI research in the early seventies. The feature that made my implementation efficient at analyzing the game's options was the data structure, essentially a conceptual geometry, used to represent and navigate possible outcomes. Unfortunately, in those days computer time was measured in real dollars, which meant you weren't allocated very much of it.

19. Additional reflections regarding automation may be of interest:

In the beginning of the chapter, I described how the information revolution provided staff departments in most corporations with what they needed to free themselves of the data monopoly imposed upon them by their DP department. As a result of that revolution, access to data became more decentralized even when the raw data was not.

In many legacy corporations in the private sector, the DP department is no longer in the business of developing software applications. DP's role has been reduced to maintaining the data center and networks responsible for warehousing corporate data, servicing personal computers, and marshalling shared applications developed outside of the corporation. The information revolution resulted in a net shift of software development out of DP and into staff departments DP supported.

Unfortunately for many programmers in DP, the staff departments they supported were looking for a different skill set. Staff departments were

chap 4 cont fn 19

typically looking for less expensive employees with the right combination of business and programming skills to quickly translate vaguely articulated business requests into some form of business model or report–output that in most cases was subject to a short shelf life. Business analysts, each armed with a personal computer, became the technology investment to replace high priced mainframe programmers in a typical DP department. Technology employed as a combination of man and machine doesn't change the reality that it was a form of automation. The intent was to produce more access to information faster and cheaper than what the DP department was charging, which was usually quite expensive.

The reallocation of resources from DP to staff departments was hard fought in a politically toxic environment with all of the sabotage, back stabbing, and palace intrigue of a soap opera. If you weren't careful–even if you were careful–you could find yourself in the midst of the conflict just by virtue of the department you happened to be working in. There was nothing professional about it because upper management was typically content to let everyone duke it out, and in most cases managers were more than happy to break the stranglehold DP seemed to have on budget allocations. (Keep in mind that until the demise of one of the largest accounting firms in the US in 2001 and accounting reforms that followed, it wasn't unheard of for a large corporation to saddle their DP with a substantial project under development by the consulting arm of the same accounting firm responsible for corporate auditing–a potential conflict of interest. Some of those projects had no real completion dates or deliverables to gauge progress.)

Once staff departments had access to what they considered 'their data', you began to see the gradual commoditization of the business analyst as the level of computer skills necessary to satisfy most requests became less demanding. The irony is that around the same time countless intractable

Notes

systems were created by staff departments and their business analysts with insufficient computer science skills to warrant some of their more ambitious projects. Since staff departments refused to ask the DP department for help–talent that could help probably wasn't there any more–an entire cottage industry of expensive consultants and contract programmers emerged to either bail a department out or take the blame.

Is what I've just described really that different from the kind of automation taking place in manufacturing? How long do you think it's going to be before routine tasks performed by business analysts, accountants, and product managers are replaced by AI? It's already happening. The point is automation isn't new; it's been progressing in one form or another since the industrial revolution. Like it or not, there are always winners and losers. What's new is the acceleration in pace resulting from pressures of globalization and advances in technology. In the current economic environment, human labor is becoming increasingly challenged to justify its worth.

The problem is when corporations embrace automation they're typically motivated by the desire to cut overhead and production costs by eliminating labor. That strategy is consistent with their goals of maximizing profit and shareholder value. Manufacturers, in particular, aren't interested in investing in robotics or AI with the goal of enabling labor to be more innovative in an economic environment where competition for market share is intense and profit margins are low. The goal is to eliminate as much labor as possible because relatively few products require highly skilled line craftsmen or are so customized automation is of limited value. We have to come to the realization that job creation and maximizing profit don't always complement each other, particularly in some industries.

Automation and AI can improve our lives, but what happens to workers they displace? If it's not the job of corporations to concern themselves with the cost to society, then whose job is it? Letting the free market decide is efficient, but the free market never has been and never will be sympathetic or accommodating. The rise of populism and nationalism in Western Europe and the United States are in part responses to economic realities where people feel threatened by what they see unfolding around them. They see limited prospects for the future. Globalization and automation have made a lot of products cheaper, but the reality is consumers don't care about how much goods cost as long as they have the money to buy them.

So, after forty years of hype, AI appears poised to make real contributions; however, it's not clear we have an answer for the threat of AI in the global market economy we currently embrace where everything that can be automated is being automated. If AI becomes a threat, it will be because we've made it a threat. If we limit our focus to using automation and AI to increase profitability in the private sector at the expense of human labor, we're probably going to have a problem. The question to be pondered regarding automation, AI, and artificial consciousness is where can they take us; what will they allow us to achieve tomorrow we only dream of today? We want those advances to take us to a tomorrow where the reasons for living are still as precious as life itself–that takes planning and foresight.

(At the risk of belaboring the point, allow me to summarize the above from a different perspective. The very thing I feared over thirty years ago appears to be approaching critical mass. The economic contract with capitalism formed during the industrial revolution, a contract where workers invested in some skill early in their careers that allowed them to find employment capable of sustaining them until retirement, is no longer working for a substantial swathe of the middle class in Western Europe and the United States. The obvious reason is those economies have evolved beyond their industrial revolutions while segments of their populations, primarily due to

chap 4 end fn 19

age and skills, haven't made the transition. The problem shouldn't surprise us because it didn't happen overnight. We can attempt to hold off pressures of globalization and automation in the U.S. by turning inward and retreating from the world stage, but it's a retreat that will ultimately be at the expense of access to markets. History has demonstrated (Organski and Kugler 1980) that conflict over access to markets is the road to war.

Interestingly enough, at the time of this writing, the U.S. economy is *statistically* returning to full employment. You might think it's a statistic that contradicts the apprehensions expressed above, yet one senses things aren't quite the same. A tectonic shift in prospects for upward mobility has taken place, too many of us are one medical emergency away from financial ruin, and wage growth in what's left of small town America is still fairly flat. The number of people hired into positions with no long-term career potential or working two jobs to make ends meet is concerning–not to mention those who dropped out of the workforce. When employers no longer treat employees as investments, they can fire them as quickly as they hire them. Employers are asking for very specific skills and experience for a reason: why invest in training an employee or invest in continuing education if you can find someone in debt up to their eyeballs for their own education and training?

The real problem going forward will be the pressure to replace or retire assets, including human resources, no longer deemed economically viable in a more competitive environment that automation and AI will invariably impact. The trick is to know your environment–your economy, your industry, your company, your employment situation–and where you fit in the unfolding history of that environment. In the U.S. we have a habit, particularly among males, of defining ourselves by what we do to make a living. You might want to rethink that habit. A machine can replace your job, but it can only replace *you* if you've equated the two.)

Epilogue

The quest for artificial consciousness offers a number of fascinating detours. In these pages we've presented a related discovery of general interest, and we've presented it with enough background and insight to allow you to decide for yourself whether you're reading fact or fiction.

The story the pyramids tell through the map they encode makes a compelling case for an extraterrestrial message; however, accepting the idea that extraterrestrials visited Earth in our ancient past is a big step. We can take that step by following the map. The more knowledge about the order of things the map reveals, the more likely it becomes that the map was designed by extraterrestrials.

We can follow the map in two directions. We can explore the psychological, phenomenological, and ontological interpretations of the generic situation the map presents to discover what those interpretations reveal about the psyche, consciousness, and being; or we can dig deeper into details of the conceptual geometry the map is built upon to discover what the conceptual geometry reveals about spatiotemporal multidimensionality, fundamental patterns, and the generic situation.

When we reflect on how all of the above come together in the astonishing way they do, it's hard to avoid feeling a bit mesmerized; nevertheless, we owe it to the designers of the map and ourselves to thoroughly understand what they were trying to tell us.

www.ingramcontent.com/pod-product-compliance
Lightning Source LLC
Chambersburg PA
CBHW080848020526
44118CB00037B/2313